Leckie
the education publisher
for Scotland

National 5
BIOLOGY
For SQA 2019 and beyond

Practice Workbook

© 2020 Leckie

001/19112020

10 9 8 7 6 5 4 3 2 1

ISBN 9780008446772

Published by
Leckie
An imprint of HarperCollins*Publishers*
Westerhill Road, Bishopbriggs, Glasgow, G64 2QT
T: 0844 576 8126   F: 0844 576 8131
leckiescotland@harpercollins.co.uk   www.leckiescotland.co.uk

This material has previously been published in the following titles:
9780008263577 *National 5 Biology Practice Question Book* by John Di Mambro and Stuart White
9780008281670 *National 5 Biology Practice Papers* by Billy Dickson and Graham Moffat

Publisher: Sarah Mitchell
Project Managers: Harley Griffiths, Lauren Murray and Fiona Watson

Special thanks to
QBS (layout);
Lauren Reid (editing);
Dylan Hamilton (proofread)

Printed and bound by CPI Group (UK) Ltd, Croydon CR0 4YY

A CIP Catalogue record for this book is available from the British Library.

Acknowledgements
Whilst every effort has been made to trace the copyright holders, in cases where this has been unsuccessful, or if any have inadvertently been overlooked, the Publishers would gladly receive any information enabling them to rectify any error or omission at the first opportunity.

ebook

To access the ebook version of this Practice Workbook visit
www.collins.co.uk/ebooks
and follow the step-by-step instructions.

ANSWERS Check your answers online:
www.collins.co.uk/pages/Scottish-curriculum-free-resources

About this book

This Practice Workbook has been designed to help you feel confident about your knowledge, and about exams and assessments. It is presented in two parts to provide maximum support in both understanding and exam experience.

The topic practice section contains lots of graded practice in every single topic you will meet on your course. You can use it to consolidate your learning at any point, and to revise and refresh your knowledge in the run-up to exam time. The questions get gradually more challenging to support and extend your knowledge at the same time.

The mixed practice section then gives you the chance to put that knowledge to use in a format and standard that reflects your exams. If you get stuck on a question, you can review the relevant topic section and then come back to try it again.

Good luck!

the education publisher
for Scotland

National 5
BIOLOGY

For SQA 2019 and beyond

Topic Question Practice

John Di Mambro, Stuart White

1 Cell structure

Exercise 1A Cell ultrastructure and functions

1 Which of the following pairs of organelles are found in both animal and plant cells?

   A Chloroplasts and permanent vacuole   **B** Cytoplasm and cell wall

   C Cell membrane and nucleus   **D** Vacuole and nucleus

> **Hint** Organelles are the small sub-cellular components that have specialised functions.

2 State which part of a nerve cell carries genetic information.

> **Hint** 'State' is an example of a 'command' word. Like 'name', 'identify' and 'list', a one-word answer or a list is what is looked for here.

> **Hint** Think generally where the genetic material is in any cell.

3 State the property of the cell membrane that controls what can enter and leave a cell.

4 Name the structures in an animal cell that synthesise the enzyme catalase.

> **Hint** Think what enzymes are made up of.

5 State the important process carried out inside a mitochondrion.

> **Hint** It's important to link this process to the need for oxygen to allow it to function.

6 Name a cell that contains a relatively high number of mitochondrion and explain why this is the case.

7 State where in a white blood cell most of the cell's biochemical reactions take place.

8 The diagram opposite shows a typical green plant cell.

   Which of the structures labelled would also be found in a muscle cell?

   A W and X   **B** X and Y

   C W and Z   **D** W, X and Y

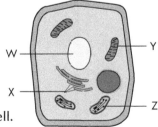

9 State where protein synthesis is carried out in a leaf palisade cell.

10 Which of the following structures would be too small to be seen with a normal light microscope?

   A Chloroplast   **B** Nucleus

   C Vacuole   **D** Ribosome

11 State the function of the chloroplast.

12 The diagram opposite shows a cell from a potato.

a Identify the parts labelled X and Y.

b State the function of the part labelled Z.

c Which organelles, found in a typical green plant cell, are missing from this cell? Explain why they are missing.

13 Name a fungus which consists of only one cell.

14 State two ways in which a fungal cell is different from a bacterial cell.

15 Which of the following statements is true?

A Fungal cells contain chloroplasts   B Fungi can synthesise their own food

C Yeast cell walls contain cellulose   D Yeast cells contain ribosomes

16 Identify the structures in the diagram below of a yeast cell.

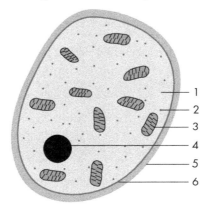

1
2
3
4
5
6

17 State one function of a plasmid in a bacterial cell.

> **Hint** While plasmids are useful in genetic engineering, that is not their natural function in a cell, so consider other options.

18 State two ways in which a fungal cell is similar to a bacterial cell.

19 Which of the following cells has a circular chromosome?

A Plant   B Animal

C Bacterial   D Fungal

20 Draw a diagram of a typical bacterial cell and identify the main structures of the cell.

21 Which two structures are common to both plant and bacterial cells?

A Cytoplasm and cell wall   B Chloroplasts and ribosomes

C Permanent vacuole and cytoplasm   D Circular chromosomes and ribosomes

22 State two ways in which a fungal cell is similar to an animal cell.

23 The following diagram shows cells from a plant and an animal.

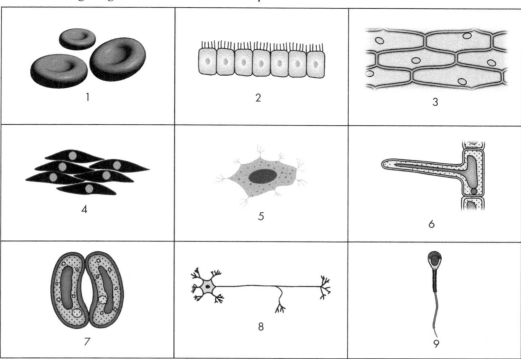

a Which of the following are all animal cells?

  A 1, 2 and 7   **B** 4, 5 and 9

  C 2, 3 and 6   **D** 7, 8 and 9

b Which of the following are all plant cells?

  A 1, 2 and 8   **B** 3, 6 and 7

  C 2, 3 and 9   **D** 2, 5, 8 and 9

c Identify a cell that would have cellulose present and the part of the cell that contains cellulose by drawing the cell and labelling that part C.

24 The following diagram shows a one-celled organism called *Euglena*.

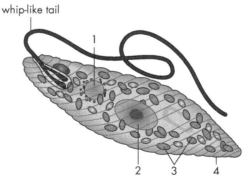

a State one reason why *Euglena* could be thought of as a plant cell.

b State one reason why *Euglena* could be thought of as an animal cell.

c State two structures, labelled on the diagram, that are shared by both animal and plant cells.

d State one structure, not labelled on the diagram, that is shared by both animal and plant cells.

25 Chloroplasts are absent from the cells of which of the following plant structures?

    A Newly formed bud and root   **B** Guard cells and phloem

    C Leaf and xylem   **D** Root and phloem

26 The diagram opposite shows a typical green plant cell.

    a A nerve cell has

        A 3, 4 and 5   **B** 1, 5 and 6

        C 2, 4 and 5   **D** 4, 5 and 6

    b Which structure would a red blood cell not have?

        A 1   **B** 3

        C 4   **D** 6

    c Which structure would an onion cell not have?

        A 1   **B** 3

        C 5   **D** 6

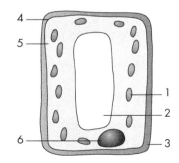

27 The table below describes some features of cells.

Decide if each statement is **TRUE** or **FALSE** then tick the appropriate box.

If the statement is **FALSE**, write the correct word(s) into the correction box to replace the word(s) <u>underlined</u> in the statement.

Statement	True	False	Correction
Bacteria lack a <u>nucleus</u>.			
Chloroplasts are found in <u>plant</u> cells.			
Ribosomes are found in <u>some</u> cells.			

Exercise 1B Cell wall

1 Which of the following is not found in animal cells but is found in plant cells?

    A Cell wall   **B** Cytoplasm

    C Cell membrane   **D** Nucleus

2 Name the main chemical component of plant cell walls.

3 The diagram opposite shows a typical green plant cell.

    a Which of the numbered structures contains cellulose?

        A 2   **B** 3

        B 4   **D** 6

    b Which of the numbered structures is the site of photosynthesis?

        A 1   **B** 3

        C 5   **D** 6

4 Write out this sentence correctly by choosing the option in the brackets that makes the sentence correct.

The walls of both bacterial and plant cells are [the same/different].

5 With reference to the cell wall, state one difference between a bacterial cell and a cheek-lining cell.

Exercise 1C Key terms

Link each term below with the correct description.

1 vacuole — **A** regulates what can enter or leave a cell

2 cell wall — **B** layer outside a plant cell that supports the cell

3 mitochondrion — **C** unit of measurement equal to 0.001 mm

4 chloroplast — **D** small organelle where aerobic respiration takes place

5 cell membrane — **E** general term for specialised sub-cellular structure

6 cytoplasm — **F** small organelle where protein synthesis takes place

7 nucleus — **G** circular piece of DNA found in bacterial cells

8 ribosome — **H** where all the cell's chemistry takes place

9 plasmid — **I** small organelle where photosynthesis takes place

10 micrometre — **J** controls all the cell's activities

11 organelle — **K** main chemical component of plant cell walls

12 cellulose — **L** fluid-filled space in plant cells that helps maintain cell shape

13 ultrastructure — **M** organism with no chlorophyll often feeding on dead animals or plants

14 yeast — **N** thread-like structure composed of DNA

15 fungus — **O** one-celled organism with no nucleus but definite cell wall and contains plasmids

16 bacterium — **P** one-celled fungus

17 chromosome — **Q** fine detail of cell structure

Exercise 1D Skills of scientific inquiry

1 The total magnification of a microscope can be calculated by multiplying the power of the eyepiece and the power of the objective lens.

   a A microscope magnifies ×100. If the eyepiece lens is ×20, calculate the power of the objective lens.

   b Using an objective lens of ×200, state what eyepiece lens would give an overall magnification of ×1000.

2 1 mm = 1000 μm.

   The average size of some bacterial cells was found to be 2.5 μm. Calculate the actual average size of these cells in mm.

   > **Hint** Practise often the conversion of micrometres to millimetres and the other way round as well. Repeated practice of different types of calculation will be of great benefit.

3 The average sizes of some cells in mm are shown in the table below.

   a Convert these to μm then draw a bar chart to show these calculated average sizes.

Type of cell	Average size (mm)	Average size (μm)
Fungal	0.008	
Animal	0.025	
Plant	0.050	
Bacterial	0.004	

   b State the ratio of the average size (mm) of bacterial cells to fungal cells.

   > **Hint** Always try to give ratios as the simplest whole numbers and in the order asked in the question.

   c State how many times smaller the average size of animal cells is compared to the average size of plant cells.

4 The diagram opposite shows some cells viewed under a microscope.

   a The average length of the cells was estimated at 0.4 mm.

   Calculate the diameter of the field of view.

   b If the field of view had been 3.00 mm, calculate the average length of the cells in mm.

   > **Hint** Remember to read the question carefully!

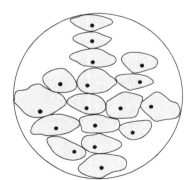

1 Cell structure

5 The approximate percentage of different types of blood cells, identified by the letters A, B, C, D, E and F, present in the blood of three different organisms is shown in the table below.

Blood cell type	Approximate percentage present		
	Human	Dog	Cat
A	60	65	50
B	30	20	30
C	2.0	5.0	5.0
D	0.5	0.5	0.5
E	6.0	7.0	7.0
F	2.5	1.5	2.0

a Express the ratio of cell types A, B and E in humans.

b State which cell type is 10 % greater in cats compared with dogs.

c Calculate the percentage of cell types A, B, D and E that make up all the blood cells in a cat.

d State which cell type is present in the same percentage as C in humans.

2 Transport across cell membranes

Exercise 2A Cell membrane

1 The diagram below shows some of the structures that make up a cell membrane.

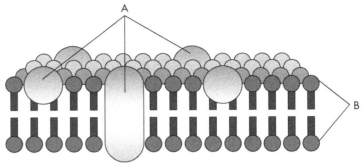

Identify the structures labelled A and B.

> **Hint** Make sure you can identify all the structural components of the cell membrane.

2 Explain what 'selectively permeable' means as applied to the cell membrane.

> **Hint** Another command word: 'Explain' usually asks you to discuss why an action has been taken or an outcome reached, the reasons and/or processes behind this action or outcome. Students often confuse explain with describe!

Exercise 2B Passive transport – osmosis and diffusion

1 State which of the following molecules can pass easily through the cell membrane.

starch – water – oxygen – carbon dioxide – protein

2 Which of the following are passive processes?

A Osmosis and active transport   **B** Diffusion and osmosis

C Active transport and osmosis   **D** Diffusion and active transport

3 Write out this sentence correctly by choosing the option in the brackets that makes the sentence correct.

In passive transport, molecules move [down/up] the concentration gradient and [do/do not] require energy.

4 Using the word bank provided, complete the following sentences by writing in the appropriate word(s).

A word may be used once, more than once, or not at all.

active – high concentration – low concentration – passive – equal concentration

Diffusion of oxygen into the blood from the lungs is an example of _____ transport in which molecules of oxygen travel from an area of _____ _____ to an area of _____ _____.

5 The following experiment was set up.

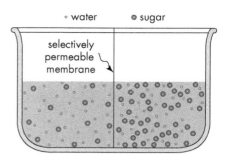

° water   ° sugar

selectively permeable membrane

a Predict what will happen to the level of the water on either side of the membrane after one hour.

> **Hint** Another command word: 'Predict' means work out what will happen.

b State the term that describes the movement of water demonstrated by this experiment.

> **Hint** Remember that osmosis is a special case of diffusion of water molecules across a selectively permeable membrane from an area of high to low water concentration.

6 Use all of the terms in the word bank given to describe what happens when a red blood cell is placed in:

a pure water

b a strong salt solution

> high water concentration – low water concentration – concentration gradient – selectively permeable membrane – burst – shrink

7 An onion cell does not burst when placed in pure water because it has a

A cell membrane   **B** nucleus

C permanent vacuole   **D** cell wall.

8 Which of the following correctly shows what happens when cells are placed in pure water?

	Plant cell	Animal cell
A	swells and bursts	shrinks
B	becomes turgid	swells and bursts
C	remains unchanged	becomes plasmolysed
D	becomes plasmolysed	remains the same

> **Hint** Always think in terms of the concentration gradient of the water molecules to answer this type of question.

9 The following diagram shows an alveolus in the lungs.

a State what X and Y each represent in the diagram.

b State the term that describes the movement of both X and Y out of and into the blood.

c Explain why X and Y move in these directions.

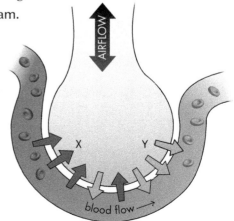

AIRFLOW

X   Y

blood flow →

10 The following diagrams show plant cells in different solutions of salt.

 A   B   C

Identify which cell was placed in each of the following solutions and explain your answer.

1 Very dilute   **2** Very concentrated   **3** Equal to the internal concentration of the cell

11 A piece of carrot was weighed before being placed in a strong sugar solution for 45 minutes. After this time, it was reweighed.

a Predict what would happen to the mass of the carrot.

Explain your answer.

b State a precaution that must be taken before reweighing the piece of carrot to make the results valid.

c Describe how the experimental results could be made more reliable.

12 The diagram below shows an experiment involving potato tissue.

A shows the potato at the start of the experiment and B shows the potato at the end of the experiment.

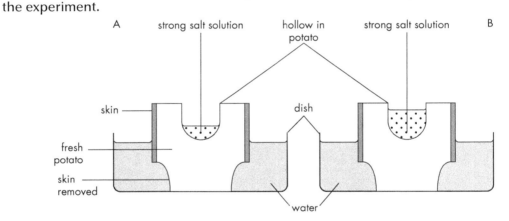

a State the process that was responsible for the change in the water levels by the end of the experiment.

b Suggest why the skin was partially removed at the start of the experiment.

c If the water had been replaced with a dilute salt solution, predict the effect this would have had on the rate of change in the water level.

Explain your answer.

d Suggest a suitable control for this experiment.

> **Hint** Another command word: 'Suggest' asks you to go beyond a one-word answer or list. Perhaps you give a proposal or an idea.

Exercise 2C Active transport

1 Define active transport.

> **Hint** Another command word: 'Define' means give the meaning.

2 State three ways in which active transport is different from diffusion.

3 State one example of active transport in animal cells and one example in plant cells.

4 The diagram below shows different ways molecules may move into and out of a cell. Which letter shows an active uptake of the molecules into the cell?

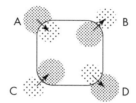

5 The diagram below shows two different processes by which molecules can be transported across the cell membrane.

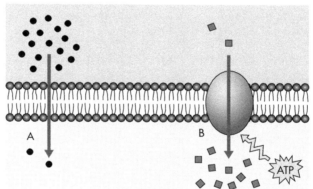

a State the name of each process.

b Predict which of these processes would not be affected by a poison that inhibits aerobic respiration.

6 The table below shows some statements describing transport across cell membranes.

Decide if each statement is **TRUE** or **FALSE** then tick the appropriate box.

If the statement is **FALSE**, write the correct word(s) into the correction box to replace the word(s) underlined in the statement.

Statement	True	False	Correction
Active transport <u>does not</u> require energy.			
Active transport moves ions from a <u>high to low</u> concentration.			
Diffusion works in the <u>same</u> direction as active transport.			

Exercise 2D Key terms

Link each term below with the correct description.

1. concentration gradient   **A** an example of passive transport

2. osmosis   **B** condition of red blood cell placed in water

3. diffusion   **C** movement of molecules or ions from a region of low to high concentration

4. selectively permeable   **D** condition of a plant cell swollen with water

5. active transport   **E** condition of a plant cell that has lost a lot of water

6. plasmolysed   **F** molecule that forms part of a cell membrane

7. turgid   **G** difference in concentration between two areas

8. phospholipid   **H** allows some but not all substances to pass through

9. burst   **I** movement of water across a selectively permeable membrane from an area of high to low concentration

Exercise 2E Skills of scientific inquiry

1. A student investigated the effect of osmosis on the diameter of discs cut from a courgette.

   The method used was as follows:

   1. The courgette was peeled.
   2. Six discs were cut from the courgette, each exactly 3 mm thick as measured by a ruler.
   3. The diameter of each disc was checked with a ruler to ensure each was exactly 12.00 mm.
   4. Five discs were each placed in a separate dish.
   5. 100 cm^3 of different concentrations of salt solution were poured into four of the dishes.
   6. 100 cm^3 of water was poured into a fifth dish.
   7. The experiment was allowed to run for 40 minutes.

8 After 40 minutes, the diameter of each of the five discs was measured.

The results of the student's investigation are shown below.

Solution	Concentration of salt (%)	Diameter of disc (mm)	
		At beginning of investigation	After 40 minutes
1	0.0	12.0	15.5
2	0.5	12.0	14.5
3	1.0	12.0	13.0
4	1.5	12.0	12.0
5	2.0	12.0	10.5

Hint Remember, in biological experiments, usually only one variable is altered at a time with all the others being controlled.

a In this investigation, state which variable the student was changing.

b State two variables that had to be kept the same throughout the investigation.

c State which of the solutions had a water concentration lower than that of the courgette.

Explain your answer.

d State which solution produced the greatest change in diameter.

e Explain why the disc was placed in water in dish 1.

f State two reasons why this investigation may not have produced reliable results.

g The sixth disc was placed in a salt solution whose strength was not known.

The diameter was 12.0 mm at the start and 10.0 mm after 40 minutes.

Which of the following is the likely concentration range for this unknown salt solution?

A 0 – 0.5 %   **B** 0.5 – 1.0 %

C 1.5 – 2.0 %   **D** 2.0 – 2.5 %

h Describe how the student could have improved on the accuracy of measuring the diameter of the cut discs.

Hint When thinking of accuracy, think of apparatus.

3 DNA and the production of proteins

Exercise 3A DNA structure and function

1 Which of the following do not contain DNA?

A Nucleus and chromosome   B Cell membrane and cell wall

C Chromosome and gene   D Nucleus and gene

2 Using the word bank provided, complete the following sentences by writing in the appropriate word(s).

A word may be used once, more than once, or not at all.

> *amino acids – complementary – bases – helix – thymine – code – messenger – proteins*

DNA is a double-stranded _____ that is held together by _____ pairs of _____. DNA carries the information needed to make _____. Adenine, guanine, cytosine and _____ make up the genetic _____. _____ RNA is a molecule that carries information from DNA to a ribosome.

3 The table below shows some of the features of DNA.

Decide if each statement is **TRUE** or **FALSE** then tick the appropriate box.

If the statement is **FALSE**, write the correct word(s) into the correction box to replace the word(s) <u>underlined</u> in the statement.

Statement	True	False	Correction
DNA is mainly found in the <u>cytoplasm.</u>			
<u>Genes</u> are sections of DNA.			

4 In a DNA molecule, guanine pairs up with

A guanine   B cytosine

C adenine   D thymine

5 a State the complementary base for thymine.

b State how many strands of DNA are in the double helix.

c State how genetic information is stored in the DNA molecule.

6 The diagram opposite is a representation of part of a DNA molecule.

a Complete the diagram by writing all the missing bases onto the diagram and labelling one phosphate group and one deoxyribose molecule.

b Identify a feature of the DNA molecule that is not represented by this diagram.

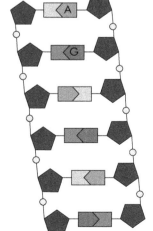

7 Which of the following diagrams best shows the four different types of bases found in DNA?

8 Using the word bank provided, complete the following sentences by writing in the appropriate word(s).

A word may be used once, more than once, or not at all.

bases – amino acids – DNA – protein – carbohydrate

Chromosomes contain the large molecule _____. The sequence of _____ is a genetic code for a sequence of _____ _____ which are joined together to form a _____.

> **Hint** The sequence of bases in a DNA molecule forms the genetic code.

Exercise 3B Protein synthesis

1 The table below shows some statements related to protein synthesis.

Decide if each statement is **TRUE** or **FALSE** then tick the appropriate box.

If the statement is **FALSE**, write the correct word(s) into the correction box to replace the word(s) <u>underlined</u> in the statement.

Statement	True	False	Correction
mRNA is made in the <u>cytoplasm</u>.			
<u>Ribosomes</u> are the organelles where protein synthesis takes place.			

2 The following diagram shows four different amino acids, A, B, C and D and the sequence of DNA bases that codes for each one.

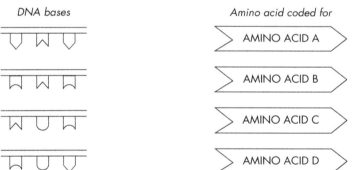

a Write out the sequence of amino acids that the following DNA sequence would code for.

b Using the DNA base shapes shown, draw the DNA sequence which would code for the following chain of amino acids.

A – A – D – B – C

3 The orders of some bases on a molecule of mRNA that code for three different amino acids, indicated by AA1, AA2 and AA3, are shown in the table below.

Order of bases on mRNA	Amino acid coded
AGA	AA1
ACA	AA2
CGC	AA3

a State the order of the bases on a mRNA that would make up a section of a protein consisting of the following sequence of amino acids.

AA1 – AA3 – AA2 – AA1

b State the order of the bases on the DNA section that coded for this sequence of amino acids.

4 Which of the following animal cells would have the largest number of ribosomes in proportion to the size of cell?

A Red blood cell   B Salivary gland cell

C Sperm cell   D Skin cell

5 The following statements refer to the stages in the formation of an enzyme, such as catalase, using information encoded in DNA.

A In cytoplasm a chain of amino acids is formed and folds to become the enzyme catalase.

B In cytoplasm amino acids are assembled to form a chain.

C mRNA molecule carries code for catalase synthesis from nucleus to ribosome.

D DNA molecule unwinds to expose the gene that codes for catalase to be copied to mRNA.

a Write these events in the correct sequence, starting with the earliest.

b State how the structure of catalase is different from other enzymes.

6 The diagram opposite represents part of the process involved in synthesising mRNA.

a State the name used to describe the pairing between bases such as cytosine and guanine, and thymine and adenine, in the DNA molecule.

b Identify a feature of the DNA molecule not shown in this diagram.

c State where this process takes place in a cell.

d After the mRNA is synthesised, state where newly synthesised mRNA travels to.

e State the organelle to which mRNA attaches.

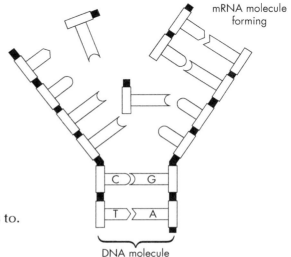

mRNA molecule forming

DNA molecule

Exercise 3C Key terms

Link each term below with the correct description.

Hint Flash cards are a powerful learning tool and great for revision too! You can make these easily using free online resources to match key terms and definitions.

1	gene	**A**	base that pairs with cytosine
2	double-stranded helix	**B**	section of DNA which codes for a protein
3	complementary base pairing	**C**	basic building unit of a protein molecule
4	adenine	**D**	molecule that carries complementary copy of a DNA section
5	mRNA	**E**	base that pairs with thymine
6	amino acid	**F**	shape of the DNA molecule
7	guanine	**G**	base that pairs with guanine
8	cytosine	**H**	each base can only be found opposite one other base
9	deoxyribonucleic acid (DNA)	**I**	nitrogen-containing chemical forming part of DNA
10	deoxyribose	**J**	chemical that can speed up a reaction
11	thymine	**K**	sugar found in DNA
12	base	**L**	base which pairs with adenine
13	enzyme	**M**	protein causing a reaction in living cell to speed up
14	catalyst	**N**	sequence of bases on DNA which codes for a protein
15	genetic code	**O**	nucleic acid which stores genetic information

Exercise 3D Skills of scientific inquiry

1 The following table shows the relative proportion of bases in an animal cell:

Base	Relative percentage
cytosine	21
adenine	32

State the relative percentage of the base guanine.

Hint Always be able to link up the complementary base pairs correctly. Make up an easy way to remember such as **T**ea **A**nd **C**ake **G**o together.

2 A length of single-stranded DNA has 500 bases.

a Complete the table opposite to show the number of thymine and cytosine bases and calculate the percentages of the bases.

b Express the ratio of adenine to guanine in this length of DNA.

c A different length of single-stranded DNA has 1000 bases. If 20 % of these bases are cytosine, calculate the numbers of each of the four bases.

Base	Number of bases	Percentage of bases
C		30
G	150	
T		
A	100	

3 The table opposite shows the mass of some components of a bacterial cell expressed as a percentage of the total mass of the cell.

a Draw a bar chart to show these data.

Component	Percentage of total cell mass
DNA	2
RNA	26
Ribosomes	30

> **Hint** Work on your graph-drawing skills. Bar charts must have equally spaced bars all the same width. Use a ruler to draw the bars. Make sure the scale uses up more than 50 % of the grid. Ensure both axes are properly labelled; and units, where these apply, must be included. Use a ruler and sharp pencil to draw the bars!

b Calculate the ratio of the percentage of RNA to the percentage of DNA.

> **Hint** Remember to give ratios as the simplest whole numbers and also in the order asked; here the RNA comes first.

4 Calculate how many cytosine molecules are present in a single-stranded section of a DNA molecule of 1000 bases if 40 % of the bases are adenine.

5 The following table shows the rate of amino acid assembly (amino acids/second) during protein synthesis in four different organisms: W, X, Y and Z.

Organism	Rate of amino acid assembly (aa/second)
W	30
X	10
Y	8
Z	4

a Complete this bar graph to show these data.

b Calculate the average rate of amino acid assembly.

c Identify which organism assembles amino acids 25 % faster than organism Y.

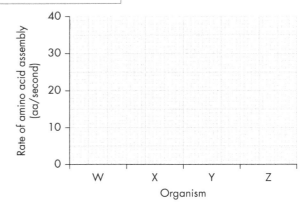

4 Proteins

Exercise 4A Protein structure and function

1 Using the word bank provided, complete the following sentence by writing in the appropriate word(s).

A word may be used once, more than once, or not at all.

amino acids – functions – sizes – base pairs

The variety of protein shapes and _____ arises from the sequence of the _____ of which the proteins are made.

> **Hint** Each protein has its own unique sequence of building blocks.

2 State two functions of proteins.

3 The following proteins are found in humans:

1 hormone   **2** amylase   **3** antibody

Which of the following correctly identifies their functions?

	Function		
	breaks down starch	acts as a chemical messenger	destroys pathogens
A	1	2	3
B	2	1	3
C	3	2	1
D	2	2	1

4 What are the basic building blocks of proteins are called?

A Bases   **B** Amino acids   **C** Sugars   **D** Fatty acids

5 Which of the following aids the body in defence against pathogens?

A Hormone   **B** Receptor   **C** Enzyme   **D** Antibody

6 Which of the following acts a chemical messenger which targets specific tissue?

A Hormone   **B** Receptor   **C** Enzyme   **D** Antibody

Exercise 4B Enzymes

1 The diagram opposite illustrates the 'lock and key hypothesis' to help explain how an enzyme works.

   a State what the key and lock each represents.

   b Identify what each of the letters A, B and C refer to.

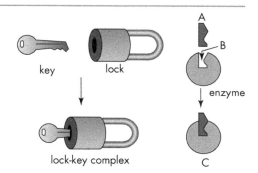

key   lock

lock-key complex

A

B

enzyme

C

2 State three properties of enzymes.

3 Biological soap powders contain enzymes to remove stains.

Explain why it is necessary for these powders to contain more than one enzyme.

> Hint   This type of question could be worth three marks so make sure your answer reflects this by giving three distinct points.

4 Explain the term 'specificity' as it applies to enzymes.

> Hint   Each enzyme has its own unique active site.

5 An enzyme may be described as a biological catalyst because it

A is made up of a chain of amino acids   B can act on any substrate

C is affected by light intensity   D can cause a reaction in a cell to speed up

6 Starch is broken down by human salivary amylase. A solution of each of these was mixed together and then the pH was adjusted to either pH 3 or pH 7 and kept at either 30° C or 50° C.

Which of the following combinations of pH and temperature would react first to produce sugar?

	pH	Temperature ° C
A	3	30
B	3	50
C	7	30
D	7	50

7 Which of the following statements about enzymes is not true?

A Enzymes act as biological catalysts

B High temperatures denature enzymes

C pH affects enzyme activity

D Enzymes are found only in cells of the digestive system

8 Complete the following sentence by underlining the correct word in the brackets.

Enzymes are [changed/unchanged] after the reaction they catalyse is complete.

9 The following numbered descriptions refer to enzyme activity:

1 optimum conditions   2 highly specific   3 forms products

4 can be reused   5 acts as a catalyst   6 increases rate of chemical changes

Select the one numbered description that best matches each underlined part of the following sentence labelled A and B.

Human lipase acts <u>only on the substrate fat</u> [A] <u>to produce fatty acids and glycerol</u> [B]

10 The following reaction involves the chemical catalase.

hydrogen peroxide ----catalase----> water + oxygen

Which of the following correctly describes the substances involved in this reaction?

	enzyme	substrate	products
A	catalase	hydrogen peroxide	water and oxygen
B	hydrogen peroxide	catalase	water and oxygen
C	catalase	water and oxygen	hydrogen peroxide
D	water and oxygen	catalase	hydrogen peroxide

11 An extract of peeled potato was prepared then filtered to produce a starch-free solution of a particular enzyme that acts on glucose-1-phosphate to produce starch.

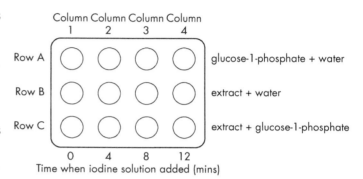

An experiment was carried out to investigate the activity of this enzyme as shown below.

Each column on the dimple tile was tested with iodine solution at four-minute intervals.

a Identify the substrate of this enzyme.

b State what iodine solution tests for.

c Predict which of the following is a likely result after adding iodine solution every four minutes [+ indicates a positive result with iodine solution].

A
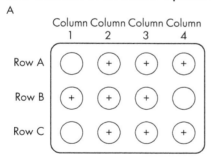

B
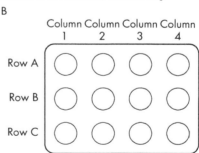

C
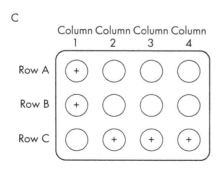

D
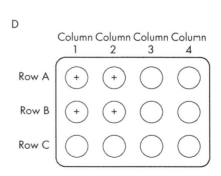

12 The graph opposite shows the effect of increasing temperature on the activity of an enzyme that can break down starch.

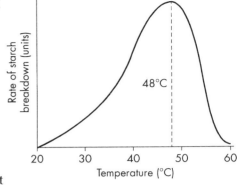

a Describe the effect of increasing temperature on the activity of the enzyme.

> **Hint** Another command word: 'Describe' means to give more detail than you would in an outline and use examples where you can.

b State the term that describes the temperature at which the enzyme is most active.

c Explain why the activity of the enzyme falls rapidly above 48°C.

d Predict the effect on the breakdown of starch if catalase had been used in this experiment.

e State one variable, other than temperature, that affects enzyme activity.

13 a State an example of an enzyme that is involved in a degradation type of reaction, stating the substrate and product(s) involved.

b State an example of an enzyme that is involved in a synthetic type of reaction, stating the substrate and product(s) involved.

c State what chemical makes up an enzyme.

14 The diagram opposite shows the activity of a plant enzyme with changing substrate concentration.

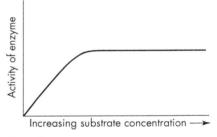

a Describe the effect of increasing substrate concentration on the activity of this enzyme.

b State two other factors that might affect the activity of this enzyme.

15 The table below shows some of the features of enzyme function.

Decide if each statement is **TRUE** or **FALSE** then tick the appropriate box.

If the statement is **FALSE**, write the correct word(s) into the correction box to replace the word(s) underlined in the statement.

Statement	True	False	Correction
Enzyme molecules are made up of chains of <u>amino acids</u>.			
Reactions in a cell go much <u>faster</u> in the presence of suitable enzymes.			
At very low temperatures enzymes are <u>denatured</u>.			

Exercise 4C Key terms

Link each term below with the correct description.

1 enzyme   **A** forms part of the cell membrane

2 antibody   **B** speeds up chemical reactions in a cell

3 structural protein   **C** acts as a chemical messenger

4 hormone   **D** binds specific molecules to the cell membrane

5 receptor   **E** protects body against disease

6 denaturation   **F** breakdown type reaction

7 degradation   **G** reaction that involves building up molecules

8 synthesis   **H** inactivation of an enzyme

9 substrate   **I** value of a factor at which an enzyme works best

10 active site   **J** end result of an enzyme-catalysed reaction

11 product   **K** chemical on which an enzyme acts

12 optimum   **L** area on enzyme where substrate binds

Exercise 4D Skills of scientific inquiry

1 An animal requires 22 amino acids to be able to manufacture the proteins it needs to function properly but it can only manufacture 11 of these through normal cell metabolism, the rest have to come from its diet.

Calculate the percentage of the animal's amino acid requirements that must be supplied in the diet.

2 The diagram below is a simplified representation of the two chains, A and B, of amino acids that form insulin in a cow.

Each amino acid is abbreviated to the first three letters of its full name.

> **Hint**   Start off by accurately logging every different amino acid in each chain.

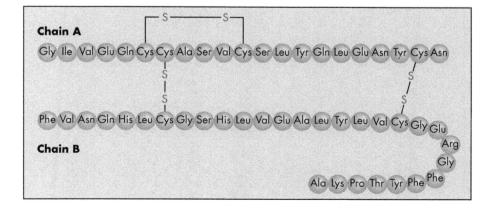

a Calculate how many more amino acids are present in chain B compared with chain A.

b Using the first three letters only, state two amino acids present in chain B but absent in chain A.

c State the ratio of Gly : Glu : Cys in both chains A and B together.

d Calculate the percentage of the amino acids in chain B of a molecule of insulin which is made up of Phe.

3 The effect of pH on the activity of two enzymes, A and B, was investigated.

The same volume of each enzyme, adjusted to the different pHs being investigated, was placed into small wells in the middle of a layer of a protein called gelatine in a Petri dish.

After 90 minutes, the activity of each enzyme was estimated by measuring the area of gelatine digested by each enzyme.

The results of the investigation are shown in the table below.

pH	Area of gelatine digested (mm²)	
	Enzyme A	Enzyme B
1	140	0
2	180	0
3	140	0
4	120	10
5	100	20
6	60	60
7	40	80
8	20	120
9	0	90
10	0	130

Hint Make sure you include the units, label each axis, join the plotted points with a ruler, point to point (not best fit), and don't plot values you don't have, such as pHs above or below the ones in the table. Make sure you use more than 50 % of the grid.

a Complete the following to show two line graphs on the same grid for these data.

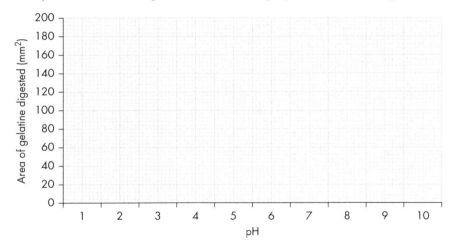

b State the pH that allowed both enzymes to work at the same activity level.

c Identify a result that looks 'unusual'.

Give a possible explanation for this unusual result.

d Explain how the results of the investigation could be made more reliable.

e Predict which enzyme is most likely to be found in the stomach, which is very acidic.

Explain your answer.

4 The graph opposite shows the activity of three human enzymes, A, B and C at different pHs.

a Describe the effect of pH on the activity of enzyme C.

b State the pH at which enzymes A and B are both working at the same percentage activity.

c State the optimum pHs for each enzyme.

d Predict the pH at which none of the enzymes will be working.

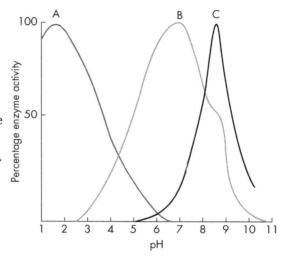

5 Genetic engineering

Exercise 5A Genetic engineering

1 Complete the following sentence by writing appropriate words into the spaces provided.

Genetic _____ can be _____ from one organism to another by genetic _____ .

2 The following diagram shows some of the stages involved in modifying a bacterial cell to make it produce a useful substance, such as insulin.

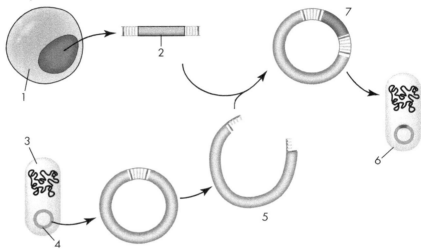

a Match each of the following with the correct number shown on the diagram.

A Human cell   **B** Plasmid   **C** Required gene

D Genetically modified bacterial cell   **E** Modified plasmid

F Bacterial cell   **G** Plasmid is opened

> **Hint**   Notice the use of 'required gene' here to specify the gene being looked for.

b State which chemicals are used in this process to cut open plasmids.

c State two advantages of producing insulin using genetic engineering to produce a chemical such as insulin.

d Suggest two reasons why the demand for insulin is likely to continue to increase.

e State one other example of a useful substance that can be produced using genetic engineering.

3 Some of the stages involved in genetic engineering are listed below.

1 Extract the required gene.

2 Identify the source chromosome that contains the required gene.

3 Insert the modified plasmid back into the original host cell.

4 Allow modified bacterial cell to multiply.

5 Insert desired gene into the bacterial plasmid.

The correct order of these stages is

A 1, 2, 3, 4, 5   **B** 4, 3, 2, 1, 5   **C** 3, 2, 1, 4, 5   **D** 2, 1, 5, 3, 4

4 Using the word bank provided, complete the following sentences by writing in the appropriate word(s).

A word may be used once, more than once, or not at all.

quickly – lab – chromosome – circular – plasmid – foreign – copies – protein – large – multiple

A _____ is a small _____ piece of DNA that is not part of the bacterial chromosome. It can make _____ of itself separately from the larger circular bacterial _____ . A _____ can carry _____ genes (genes from other organisms) cheaply and_____in the _____. They are good for making many _____ of the required gene as well as the _____ product of the gene in _____ quantities.

5 Haemophilia is a condition in humans that results in a failure to clot blood. This failure is due to a lack of a 'clotting factor'.

Genetic engineering has allowed the manufacture of this clotting factor using the following stages.

A Extracting the clotting factor.

B Growing large numbers of the genetically modified bacterial cells.

C Purifying the clotting factor.

D Identifying the required gene that codes for the clotting factor.

E Isolating the required gene that codes for the clotting factor.

F Inserting the required gene into a bacterial plasmid.

Using the letters, arrange these stages in the correct order.

6 The diagram below shows some of the structures associated with a bacterial cell.

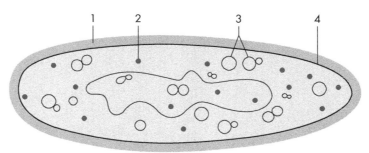

a Which labelled structure is often used in genetic engineering?

   A 1   **B** 2   **C** 3   **D** 4

b Which labelled structures have to be passed through to allow genetic engineering to take place?

   A 1 and 2

   B 2 and 3

   C 1 and 4

   D 2 and 4

7 Which of the following definitions best describes a plasmid?

A Structure in an animal cell that can be genetically programmed

B A large chromosome found in a bacterial cell

C Small ring of DNA that can act as a carrier for genes

D Small circle of DNA that carries most of a cell's genetic information

8 Two stages of genetic engineering that occur in sequence are shown below.

Plasmid is extracted from a bacterium.

Plasmid is cut open using an enzyme.

Which of the following is the next stage?

A Genetically modified bacteria multiply and produce human protein

B Section of human DNA with required gene is identified

C Required gene is cut out of the human chromosome

D Required gene is inserted into the plasmid

9 Cheeses made in the past using an enzyme, called rennin, from a calf's stomach, can now be made by using genetically modified (GM) yeast cells.

The diagram below shows some of the procedures used to reprogramme yeast cells to make rennin.

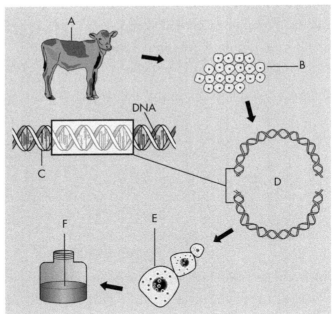

a Identify which letter refers to each of the following descriptions.

   1 Calf producing rennin.

   2 Rennin extracted from genetically modified yeast cells.

   3 Identification and removal of the gene for producing rennin.

   4 Insertion of gene into bacterial plasmid acting as a carrier.

   5 Plasmid inserted into yeast cells.

   6 Cells from calf's stomach.

b Suggest two advantages of rennin produced in this way.

10 The table below shows some statements concerning genetic engineering.

Decide if each statement is **TRUE** or **FALSE** then tick the appropriate box.

If the statement is **FALSE**, write the correct word(s) into the correction box to replace the word(s) <u>underlined</u> in the statement.

Statement	True	False	Correction
Bacteria have a large <u>circular</u> chromosome.			
Genetic engineering is the transfer of <u>whole chromosomes</u> from one species to a different species.			
<u>Plasmids</u> can be used in genetic engineering to carry genes.			

Exercise 5B Key terms

Link each term below with the correct description.

1 plasmid   **A** acts like chemical 'scissors' in genetic engineering

2 enzyme   **B** produces genetically modified organisms

3 genetic engineering   **C** useful substance that can be produced using genetic engineering

4 growth hormone   **D** small circular DNA structure found in bacterial cell

5 source chromosome   **E** changing an organism's genetic material

6 genetic modification   **F** where the required gene is located

7 gene   **G** process of genetically modifying a cell

8 species   **H** basic unit of heredity

9 transformation   **I** group of individuals which can breed to produce fertile offspring

1 The data below shows the number of patients taking a genetically engineered form of insulin in an area of Scotland.

Number of patients	Patients' age (years)
40	0–10
120	11–20
230	21–30
120	41–50
20	51–60

a Draw a bar chart to represent these data.

b Identify one trend in the number of patients taking this genetically modified insulin and the ages of patients being studied.

c Calculate the percentage decrease in the number of patients aged 51–60 years taking this modified insulin compared to those aged 21–30 years.

d State the ratio of patients taking this modified insulin aged 0–10, 21–30 and 41–50 years.

e The number of patients taking this modified insulin in the age range 61–70 fell to 18. Express this change as a percentage of the number of patients in the age range 51–60.

2 A sample of 100 people were asked their opinions on genetic engineering in 1985. Five years later, a different sample of 100 people were asked for their opinions. This was repeated until six sets of data were obtained.

Their responses were divided into three different categories, those who thought the:

1 benefits of genetic engineering far outweigh any potential harmful effects

2 benefits of genetic engineering were very similar to the potentially harmful effects

3 harmful effects of genetic engineering were far greater than the potentially beneficial effects

The graph below shows the responses as percentages of the total sample.

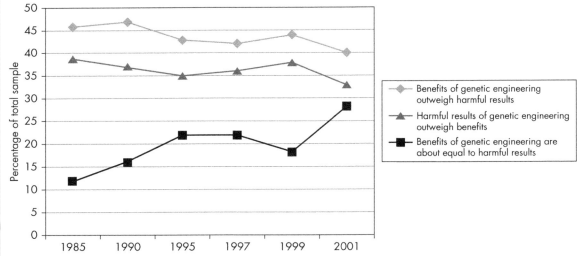

a Select one set of responses and comment on any trend over the period of the survey.

b State which set of responses produced the largest difference in 2001 compared with 1985.

c State the two years between which the percentage change was the greatest.

State the set that showed this change.

6 Respiration

Exercise 6A Stored chemical energy

1 Using the word bank provided, complete the following sentences by writing in the appropriate word(s).

A word may be used once, more than once, or not at all.

chemical – glucose – enzymes – respiration – cells – photosynthesis

The _____ energy stored in the sugar _____ needs to be released by all _____ through a series of reactions controlled by _____. This process is called _____.

Exercise 6B ATP

1 Using the word bank provided, complete the following sentences to show how energy from the breakdown of ATP can be used.

protein synthesis – division – transmission – contraction

a The _____ of muscle cells

b Allowing an increase in cell numbers by cell _____

c Carrying messages rapidly by _____ of nerve impulses

d Joining amino acids together in the process of _____ _____.

2 Which of the following would not require energy supplied by the breakdown of ATP?

A Synthesis of the enzyme catalase

B Transmission of nerve impulses

C Skeletal muscles contracting

D Passive diffusion of oxygen from leaf cells

3 Which of the following processes would require energy supplied by the breakdown of ATP?

A Breakdown of amino acids in the liver

B Digestion of protein by pepsin in the stomach

C Formation of a protein from amino acids

D Aerobic respiration of glucose

Exercise 6C Respiration and fermentation

1 Which of the following enters a white blood cell by diffusion and is broken down to release energy?

A Glucose   B Oxygen

C Water   D Starch

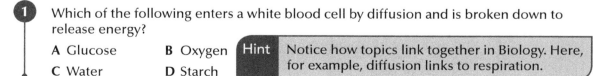

Hint Notice how topics link together in Biology. Here, for example, diffusion links to respiration.

2 Some of the stages involved in aerobic respiration follow:

1 Carbon dioxide, water and energy are released.

2 Pyruvate enters a mitochondrion.

3 Two molecules of ATP are produced.

4 Glucose is broken down to pyruvate in the cytoplasm.

5 Inside the mitochondrion, pyruvate is completely broken down.

Arrange these stages, starting with the earliest.

> **Hint** Flow charts are excellent ways of remembering complex information such as the pathway of aerobic respiration.

> **Hint** Students often confuse fermentation with anaerobic respiration.

3 The table below compares aerobic respiration and fermentation.

Write the letters of the descriptions into the correct columns.

A Requires oxygen

B Does not require oxygen

C Produces alcohol

D Large number of ATP molecules are produced

E Takes place in the mitochondria of a cell

Aerobic respiration	Fermentation

4 Fermentation by yeast produces

A oxygen, ATP and ethanol

B ethanol, carbon dioxide and water

C oxygen, carbon dioxide and ethanol

D ATP, ethanol and carbon dioxide

5 To function aerobically, the biceps muscle needs

A glucose and oxygen

B oxygen and lactate

C glucose and lactate

D glucose and carbon dioxide

> **Hint** Don't be put off by a reference to the biceps here; just consider the cells that make up the biceps.

6 Which of the following shows the end-products of fermentation in the biceps muscle?

A Ethanol and ATP   **B** Oxygen and lactate

C Lactate and carbon dioxide   **D** ATP and lactate

7 Which of the following descriptions apply to fermentation in yeast cells?

	Where it occurs	Relative number of ATP synthesised
A	cytoplasm	two
B	cytoplasm	large number
C	mitochondria	two
D	mitochondria	many

8 The diagram opposite shows that yeast cells can respire by fermentation.

a State the purpose of the glucose present.

b Suggest one reason why the glucose was boiled.

c Suggest the purpose of the layer of oil.

d The indicator changes colour from red to yellow in the presence of the gas produced during fermentation.

   State the gas that causes this change.

e The thermometer showed a slight increase in temperature.

   Suggest a possible reason for this change.

f Suggest how the reliability of the results could be improved.

g A possible control for this experiment would be using an equal mass of previously boiled yeast.

   State the purpose of this control.

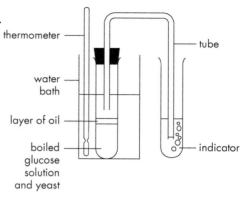

9 Which of the following stages in respiration would result in the production of the largest number of ATP molecules?

A Glucose ⟶ pyruvate

B Pyruvate ⟶ lactate

C Glucose ⟶ carbon dioxide and water

D Lactate ⟶ carbon dioxide and water

10 The diagram below shows some of the stages involved in the complete aerobic respiration of glucose.

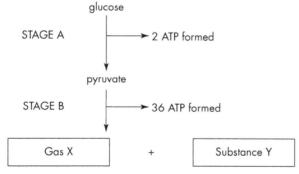

a Identify gas X and substance Y.

b Where in the cell does each of the stages, A and B, take place?

11 The following table shows some statements associated with respiration.

Decide if each statement is **TRUE** or **FALSE** then tick the appropriate box.

If the statement is **FALSE**, write the correct word(s) into the correction box to replace the word(s) underlined in the statement.

Statement	True	False	Correction
Protein is the most common source of energy in cells.			
ATP acts as a link between energy-releasing reactions and those reactions that require energy.			
The first stage of aerobic respiration takes place in the mitochondrion.			
Fermentation results in 2 ATP being released from the breakdown of glucose in the presence of oxygen.			

Exercise 6D Key terms

Link each term below with the correct description.

1 ATP   **A** energy-rich sugar that is broken down during respiration

2 glucose   **B** alcohol produced by yeast cells in the absence of oxygen

3 aerobic   **C** energy-rich molecule produced in both animal and plant cells

4 lactate   **D** requires oxygen

5 respiration   **E** chemical produced in cytoplasm of an animal cell in the absence of oxygen

6 fermentation   **F** organelle where aerobic respiration is completed

7 mitochondrion   **G** biochemical process of releasing energy from glucose

8 ethanol   **H** process which only partially breaks down pyruvate

Exercise 6E Skills of scientific inquiry

1 The following apparatus was set up to investigate the effect of temperature on the rate of oxygen uptake by live maggots.

chemical which absorbs carbon dioxide

cotton wool

live maggots

U tube containing coloured fluid

chemical which absorbs carbon dioxide

cotton wool

glass beads

U tube containing coloured fluid

The apparatus was kept in a water bath set at 5° C and left for 15 minutes with the taps open to the air.

The taps were then closed and the apparatus left for one hour.

The oxygen used up by the maggots was measured as a change in the height of the coloured liquid in the capillary tube, which was converted to a volume (cm³ per hour).

The same procedure was repeated over a range of temperatures and the results of the tube that contained the maggots are shown in the table below.

Temperature (° C)	Oxygen uptake (cm³ per hour)
5	0.20
10	0.30
15	0.40
20	0.45
25	0.50
30	0.55

a Complete the following line graph of this data.

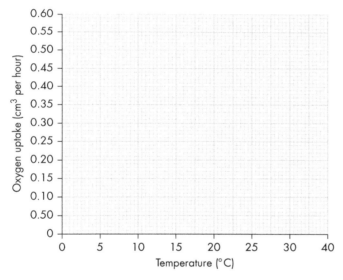

b Describe the relationship between temperature and oxygen uptake by the live maggots.

c Suggest why glass beads, equal in mass to the live maggots, were used.

d State the purpose of the cotton wool in the apparatus.

e Describe how the validity of these results could be improved.

f If the increase in oxygen uptake continued at the same rate over the range 15° C to 30° C, predict the oxygen uptake at 35° C.

2 A gardener collected waste materials, such as grass clippings, potato peelings, leaves and kitchen scraps, to form a 'compost heap' used to generate nutrients that she could apply later in the season to growing plants.

The diagram below shows the compost in section.

After several months, the temperature (°C) at each of the sample points, A to G, was recorded. These are shown in the table below.

Sample point	Temperature (° C)
A	20
B	22
C	35
D	35
E	35
F	25
G	18

a Complete the following line graph to represent this data.

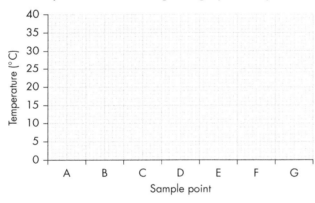

b Describe the trend in temperature readings from sample point A to sample point G.

c A number of organisms are often found in a well-established compost heap such as earthworms, aerobic bacteria and anaerobic bacteria.

Some bacteria can survive in the complete absence of oxygen.

Identify the likely sample point where such bacteria would be found and explain your answer.

7 Producing new cells

Exercise 7A Mitosis

1 The diagram below shows a cell undergoing mitosis.

 a From the diagram, identify structures X, Y and Z.

 b Describe the functions of structures Y and Z.

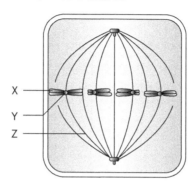

2 Put the following stages of mitosis into the correct sequence.

 A Chromosomes line up at the equator of the cell.

 B Daughter chromosomes gather at the end of the cell.

 C Chromosomes become visible as pairs of identical chromatids.

 D Spindle fibres pull chromatids to opposite poles of the cell.

> **Hint** You should be able to match diagrams with the correct sequence of the stages.

3 The diagram opposite shows four different stages of mitosis.

 a Put the stages in the diagram opposite into the correct sequence.

 b Which stage in the diagram involves chromatids being pulled apart by the spindle fibres?

 c Which is the earliest stage in the diagram that involves chromosomes becoming visible as pairs of identical chromatids?

> **Hint** You should make sure you learn the difference between a chromosome and a chromatid.

4 State three ways in which mitosis is important to living organisms.

5 Describe mitosis in terms of number of cells produced and the chromosome complement of each of these cells.

6 Which of the following involves mitosis?

 A Photosynthesis

 B Protein synthesis

 C Aerobic respiration

 D Maintenance of chromosome complement

7 The following terms refer to events in mitosis:

  A Chromosomes   **B** Spindle fibres

  C Centromere   **D** Equator

Match each letter to the statements below.

a Holds two chromatids together

b Pulls chromatids apart to opposite poles of the cell

c Is the area where chromosomes line up

8 Look at the diagram opposite.

a State the number of chromosomes present in the cell.

b State the number of chromatids present in the cell.

c State the number of chromosomes that would be present in each daughter cell produced if this cell underwent mitosis.

Hint	Daughter cells always maintain the diploid chromosome complement and so have the same number as the original cell.

Exercise 7B Stem cells

1 State one feature of a stem cell.

2 State two processes that stem cells are involved in.

3 Describe how stem cells are able to self-renew.

4 State the group of living organisms that stem cells are found in.

5 The following statements refer to stem cells. State which of the statements about stem cells are **TRUE** and which are **FALSE**. If **FALSE**, rewrite the statement to make it correct.

a Located in animals

b Involved in growth and respiration

c Have the ability to self-renew

d Are specialised cells

e Have the potential to become one type of cell

Exercise 7C Specialisation

1 State what is meant by the term specialisation.

2 State what each of the following are composed of:

a Tissues

b Organs

c Systems

3 Complete the following flow diagram to display the hierarchy in living organisms.

Cells → _____ → _____ → _____ → _____

4 State three examples of:

  a Cells   **b** Tissues   **c** Organs   **d** Systems

5 Decide if each of the following statements is **TRUE** or **FALSE** and tick the correct box. If the answer is **FALSE**, write the correct word(s) in the correction box to replace the word underlined in the statement.

Statement	True	False	Correction
Spindle fibres pull chromatids apart to opposite poles of the cell.			
Mitosis provides new cells for growth and repair of damaged cells and maintains the haploid chromosome complement.			
Stem cells in animals are specialised cells which can divide in order to self-renew.			
Specialisation of cells leads to the formation of a variety of cells, tissues and organs.			
Groups of organs which work together form organisms.			

Exercise 7D Skills of scientific inquiry

1 The table below shows the results of an experiment by students who compared two different nutrient agars on the growth of bacterial cells as measured by the number of colonies in a Petri dish.

Use the data in the table to answer the questions.

Time (hours)	Number of colonies	
	Basic nutrient agar	Nutrient-rich agar (extra glucose and amino acids)
0	0	0
6	1	2
12	2	4
18	3	7
24	4	16
30	7	27
36	6	63

a Describe the general trend in the number of colonies shown by the results using basic nutrient agar.

b Calculate the simple whole number ratio of the number of colonies in basic and nutrient rich agar at 24 hours.

c Calculate the percentage increase in the number of colonies for nutrient-rich agar between 12 and 24 hours.

d Calculate the average increase in colonies (per hour) for basic nutrient agar during the experiment.

e Predict the number of colonies for nutrient-rich agar at 42 hours, assuming rate of growth remains constant.

Exercise 7E Key terms

Link each term below with the correct description.

1	cell	**A**	when chromosomes become shorter and thicker
2	cell division	**B**	protein thread produced during mitosis to pull chromatids apart
3	centromere	**C**	cells produced at the end of mitosis
4	chromatid	**D**	process to increase the number of cells in an organism
5	chromosome complement	**E**	one half of a duplicated chromosome
6	condense	**F**	number of chromosomes possessed by the cells of a species
7	differentiation	**G**	imaginary line down the centre of the cell where chromosomes line up
8	diploid	**H**	area of the cell to which spindle fibres pull chromatids
9	equator	**I**	structure to which the chromatids of a chromosome are attached
10	ethical issues	**J**	produces two new daughter cells that are identical to each other, and to the parent cell
11	haploid	**K**	two sets of chromosomes
12	identical daughter cells	**L**	one set of chromosomes
13	mitosis	**M**	unspecialised cells in animals
14	multicellular	**N**	process by which an unspecialised cell becomes specialised
15	organ	**O**	moral questions or problems
16	system	**P**	a cell that has not yet developed into one particular type
17	organism	**Q**	group of similar tissues working together
18	pole	**R**	organism which consists of many cells
19	specialisation	**S**	unit of life
20	spindle fibre	**T**	the process whereby a cell becomes adapted to one specific function
21	stem cell	**U**	organism that consists of only one cell
22	tissue	**V**	an individual living thing / made up of many systems
23	unicellular	**W**	group of similar cells working together
24	unspecialised	**X**	group of organs working together

8   Control and communication

Exercise 8A Nervous system and reflex arc

1   Name the:

   a Two main parts of the nervous system.

   b Two main parts of the central nervous system (CNS).

   c Three main parts of the brain.

2   Describe one function of the:

   a Cerebellum

   b Cerebrum

   c Medulla.

3   Identify the following parts of the brain shown on the diagram opposite.

   a X

   b Y

   c Z

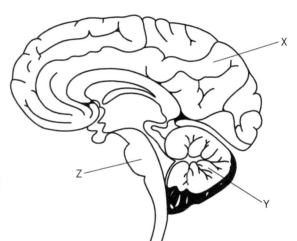

4   Which part of the brain co-ordinates muscle contraction?

   A Cerebrum   **B** Medulla

   C Cerebellum   **D** Spinal cord

5   Describe the function of a neuron.

6   **a** State the three types of neurons.

   b Describe the specific function of each neuron within a reflex arc.

7   State the function of a receptor.

8   **a** Name the gap between neurons.

   b Explain how the messages carried by neurons are able to pass over this gap.

9   State what a reflex is.

> **Hint**   You should memorise the pathway of a reflex arc – using a flow diagram is a useful way to do this.

10   Draw a flow diagram showing the pathway that an electrical impulse follows during a reflex arc.

11   State the function of a reflex.

12 Identify the following parts of a reflex arc shown on the diagram.

V

W

X

Y

Z

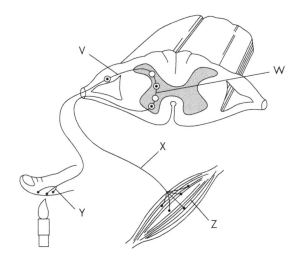

Exercise 8B Hormones and blood glucose regulation

1 State one function of an endocrine gland.

2
a State how hormones travel around the body.

b State one function of hormones.

c Explain how a hormone acts upon a target tissue within the body.

> **Hint** Students are often unable to recall the term 'target tissue'; keep this one in mind when describing hormone action.

3 Describe the roles of the following in blood glucose regulation.

a Insulin

b Glucagon

c Pancreas

d Liver

e Glycogen

4 Describe the expected response from the blood glucose regulation system in the following circumstances:

a Blood glucose level decreases below normal

b Blood glucose level increases above normal

c During/after intense exercise

d After eating a large carbohydrate meal

e During fasting

5 Name the organ where receptor cells that detect changes in blood glucose levels are found.

6 State the form in which carbohydrate is stored in the body and name the organ where this storage takes place.

7 Complete the diagram below showing blood glucose regulation.

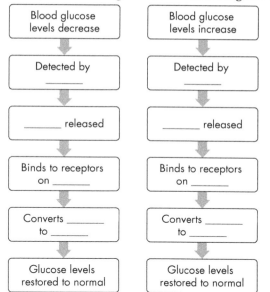

Blood glucose levels decrease	Blood glucose levels increase
↓	↓
Detected by _____	Detected by _____
↓	↓
_____ released	_____ released
↓	↓
Binds to receptors on _____	Binds to receptors on _____
↓	↓
Converts _____ to _____	Converts _____ to _____
↓	↓
Glucose levels restored to normal	Glucose levels restored to normal

Hint You should spend some time learning the different terms involved. Be careful that you don't confuse the terms glucose, glucagon and glycogen.

8 Which of the following is not a correct description of hormone function?

A Travels in blood

B Travels via electrical impulse

C Secreted by endocrine glands

D Is a chemical messenger

9 Which of the following describes the correct hormonal change to the change in blood glucose levels following exercise?

	Blood glucose change	Hormonal change
A	increased	glucagon secretion increased
B	increased	insulin secretion increased
C	decreased	glucagon secretion increased
D	decreased	insulin secretion increased

10 Decide if each of the following statements is **TRUE** or **FALSE** and tick the correct box.

If the answer is **FALSE**, write the correct word(s) in the correction box to replace the word underlined in the statement.

Statement	True	False	Correction
CNS consists of <u>nerves</u> and spinal cord.			
<u>Effectors</u> detect sensory input/stimuli.			
Chemicals transfer messages between neurons at <u>synapses</u>.			
<u>Endocrine</u> glands release hormones into the bloodstream.			
A target tissue has cells with complementary <u>receptor</u> proteins for specific hormones.			

Exercise 8C Skills of scientific inquiry

 1 Use the data below to answer the questions that follow.

The table shows the results of a glucose tolerance test on two individuals, X and Y. During the test these individuals are given glucose drinks and their blood glucose levels recorded (mmol/l), before and at certain time intervals afterwards (minutes). The individuals did not eat (fasted) overnight before this test.

Time after consuming glucose drink (mins)	Blood glucose levels (mmol/l)	
	Individual X	Individual Y
0 (fasting level)	8.8	4.1
30	24.6	8.2
60	21.2	10.6
90	18.0	8.2
120	17.6	6.6
150	16.0	5.4
180	14.4	4.8

a Describe the general trend shown by the results of individual Y's blood glucose levels.

b Calculate the simple whole number ratio of blood glucose levels of individual X and Y at 30 minutes.

c Calculate the percentage increase in blood glucose level for individual Y between 0 and 30 minutes.

d Calculate the average increase in blood glucose (per minute) of individual X during the test.

> **Hint** Average increase is calculated by dividing the actual increase by the total time of the experiment.

e Predict the blood glucose level of individual X after 210 minutes, assuming it continues to change at the same rate as 120 minutes onwards.

f When comparing these results, students calculated the percentage increase in blood glucose levels for both people rather than the actual increase. Suggest why.

g People with type 1 diabetes are unable to produce insulin. Suggest which of these two people is diabetic. Give a reason for your choice.

h Give two variables that should be kept constant during this glucose tolerance test.

2 Explain what a control experiment is.

Exercise 8D Key terms

Link each term below with the correct description.

1	central nervous system (CNS)	**A**	transfers messages across synapses
2	nervous system	**B**	tissues which bind specific hormones
3	cerebrum	**C**	sugar used as respiratory substrate in cell
4	cerebellum	**D**	storage form of glucose in liver
5	medulla	**E**	sites on the outside of cells which specific hormones bind to
6	neuron	**F**	organ which releases hormones into the blood stream
7	sensory neuron	**G**	passes the information to the central nervous system from a receptor
8	inter neuron	**H**	organ which stores glycogen
9	motor neuron	**I**	organ which detects changes in blood glucose levels. Secretes insulin and glucagon
10	receptor	**J**	nerve cell
11	electrical impulse	**K**	hormone released by pancreas when blood glucose levels are low. Converts glycogen to glucose in liver
12	synapse	**L**	hormone released by pancreas when blood glucose levels are high. Converts glucose to glycogen in liver
13	chemical transmitter	**M**	gap between neurons
14	reflex	**N**	transmits impulse from CNS to effector, enabling a response to occur, which can be a rapid action from a muscle or a slower response from a gland
15	effector	**O**	controls sensory input/stimuli
16	hormone	**P**	control of muscle co-ordination and balance
17	endocrine gland	**Q**	control of involuntary processes such as heart rate and breathing
18	target tissue	**R**	control of conscious thought, memory, speech, learning, language, reasoning, emotions
19	receptor protein	**S**	connects sensory and motor neurons in spinal cord
20	insulin	**T**	CNS and nerves
21	glucose	**U**	chemical messenger
22	glucagon	**V**	carry messages along neurons
23	glycogen	**W**	brain and spinal cord
24	pancreas	**X**	body part which responds to the electrical impulse coming from a motor neuron
25	liver	**Y**	a rapid automatic response to a stimulus (anything an organism can detect)

9 Reproduction

Exercise 9A Animal reproduction

1 State what is meant by the term diploid.

2 State what is meant by the term haploid.

> **Hint** Learn the difference between diploid and haploid, but also learn examples of both.

3 State what a gamete is.

4 Explain why gametes need to be haploid.

5 State whether each of these cells are diploid or haploid.

a Pancreas   b Sperm

c Kidney   d Pollen

e Phloem   f Ovule

g Muscle   h Skin

i Egg

6 Which of the following pairs of human cells have the same number of chromosomes?

A Pancreas and kidney   B Sperm and pancreas

C Sperm and muscle   D Egg and kidney

7 Which of the following pairs of human cells have the same number of chromosomes?

A Pancreas and egg   B Sperm and muscle

C Egg and muscle   D Egg and sperm

8 State the names of the two organs that produce gametes in animals.

9 State the site of production of:

a Sperm in male animals.

b Eggs in female animals.

10 The diploid number of chromosomes in a cell from a chicken is 78. Which of the following is correct?

	Chicken cell type	Number of chromosomes
A	Muscle	39
B	Zygote	39
C	Egg	78
D	Pancreas	78

11 Read the table below, comparing the structure of sperm and egg cells, and select the correct option from each pair.

	Sperm	Egg
Male/female	male/female	male/female
Relative size	larger/smaller	larger/smaller
Site of production	testes/ovary	testes/ovary
Tail present	yes/no	yes/no

12 Describe the process of fertilisation.

13 State the term used to describe a fertilised egg.

Exercise 9B Plant reproduction

1 State the names of the male and female gametes produced by the sex organs in plants.

2 State where in plants the following are produced:

a Pollen

b Ovules.

3 The diagram opposite shows some structures in a flowering plant.

Identify the letter which indicates the following.

a Ovary

b Anther

c Site of ovule production

d Site of pollen production

> **Hint** Both plants and animals have ovaries.

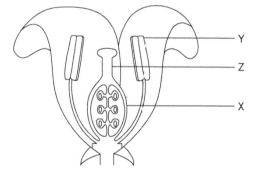

4 Which of the following pairs of plant cells has the same number of chromosomes?

A Ovule and pollen   **B** Pollen and phloem

C Xylem and ovule   **D** Xylem and pollen

5 Which of the following pairs of plant cells has the same number of chromosomes?

A Xylem and pollen   **B** Pollen and phloem

C Ovule and xylem   **D** Xylem and phloem

6 The diploid number of chromosomes from a wheat cell is 42. Which of the following is correct?

	Wheat plant cell type	Number of Chromosomes
A	Ovule	42
B	Phloem	21
C	Zygote	84
D	Pollen	21

7 Complete the table below.

Organism	Gamete	Organ of production	Haploid/Diploid
Plant		Anther	
			Haploid
Animal	Sperm		

8 Complete the table below.

Cell	Gamete?	Plant/Animal/Both	Haploid/Diploid
Pancreas	No		
Xylem			
Egg		Animal	
Pollen	Yes		
Zygote			
Sperm			Haploid
Ovule			

9 Decide if each of the following statements is **TRUE** or **FALSE** and tick the correct box.

If the answer is **FALSE**, write the correct word(s) in the correction box to replace the word underlined in the statement.

Statement	True	False	Correction
Human body cells are diploid, except gametes, which are haploid.			
Eggs are produced in the ovule.			
Fertilisation is the fusion of the nuclei of the two haploid gametes to produce a diploid egg, which divides to form an embryo.			

Hint | Remember only to replace the underlined word from the statement in the correction box, not the entire statement.

Exercise 9C Skills of scientific inquiry

1 The table opposite shows the chromosome numbers found in the cells of different organisms.

   a Select the organisms that have the same number of chromosomes in their cells.

   b State the chromosome numbers of barley and wheat as a simple whole number ratio.

   c Calculate the percentage difference between the chromosome numbers of a human and a crab-eating rat.

Organism	Chromosome number
Fruit fly	8
Chicken	78
Barley	14
Wheat plant	42
Human	46
Crab-eating rat	92
Great white shark	82
Rhesus monkey	42
Black mulberry	308
Red king crab	208

> **Hint** Percentage change is calculated by dividing the change in values by the original value and then multiplying by 100.

   d Calculate how many times greater the chromosome number of black mulberry is compared to the chromosome number of barley.

   e Calculate the average number of chromosomes of the organisms in the table.

Exercise 9D Key terms

Link each term below with the correct description.

1 gamete   **A** site of production of sperm

2 sperm   **B** site of production of pollen

3 testes   **C** site of production of eggs in animals/ovules in plants

4 egg   **D** male gamete in plants

5 ovary   **E** male gamete in animals

6 pollen   **F** haploid sex cell

7 anther   **G** fusion of the nuclei of two haploid gametes to produce a diploid zygote

8 ovule   **H** female gamete in plants

9 diploid   **I** female gamete in animals

10 haploid   **J** contains two sets of chromosomes; all cells except gametes

11 fertilisation   **K** contains one set of chromosomes; gametes only

12 zygote   **L** cell produced after fertilisation from the fusion of the gamete nuclei

10 Variation and inheritance

Exercise 10A Variation

1 State what is meant by variation.

2 Give a definition for each of these types of variation.

   a discrete

   b continuous

3 **a** Describe what is meant by the following types of inheritance.

     i single gene

     ii polygenic

   b For each type of inheritance (above), state the type of variation it is associated with.

4 For each of the following characteristics, state which type of variation it is associated with.

   a height

   b blood group

   c weight

   d fingerprint type

Exercise 10B Inheritance

1 Give a definition for each of the following genetic terms.

   a gene

   b allele

   c phenotype

   d genotype

   e dominant

   f recessive

   g homozygous

   h heterozygous

   i P

   j F_1

   k F_2

   l monohybrid cross

> **Hint** Know the difference between phenotype and genotype, as students often get them confused.

2 Explain why predicted ratios among offspring are often different from actual ratios.

> **Hint** This question appears often in examinations so it is worth paying close attention to the answer.

3 Most features of an individual phenotype are

A polygenic and show discrete variation

B controlled by single gene inheritance and show discrete variation

C polygenic and show continuous variation

D controlled by single gene inheritance and show continuous variation

4 The family tree below shows the inheritance of the ability to roll the tongue.

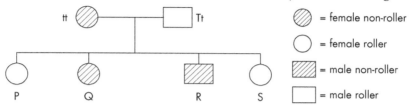

Which of the F_1 generation are homozygous?

A P and Q   B Q and R

C R and S   D P and S

5 Decide if each of the following statements is **TRUE** or **FALSE** and tick the correct box.

If the answer is **FALSE**, write the correct word(s) in the correction box to replace the word <u>underlined</u> in the statement.

Statement	True	False	Correction
The different forms of a gene are known as <u>polygenic</u>.			
<u>Discrete</u> variation tends to contain a wide range of values from one extreme to another.			
<u>Dominant</u> traits require two recessive alleles in order for the characteristic to be expressed.			
<u>Genotype</u> is the physical expression of a gene.			
When an individual has two identical alleles for a characteristic, this is known as <u>homozygous.</u>			

6 The following diagram shows the inheritance of petal colour in roses for three generations.

P Phenotype: Red   X   White

P Genotype: RR   rr

F_1 Genotype: Rr

F_2 Genotypes: RR and Rr and rr

a Which of the generations contains heterozygous individuals?

b Which of the generations contains homozygous individuals?

c Draw a Punnett square to show the parental cross and complete it to show the offspring.

d Draw a Punnett square to show the results of a F_1 self-cross.

e State the F_2 phenotypic ratio.

f State the F_2 genotypic ratio.

7 The diagrams below show the same sections of matching chromosomes found in four humans, A, B, C and D.

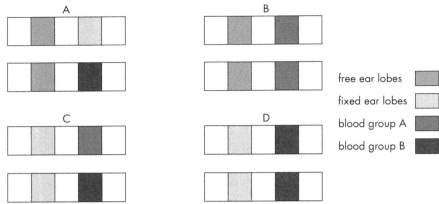

The alleles shown on the chromosomes can be identified using the above key.

State the letter(s) to which these statements apply.

a Blood group is homozygous.

b Ear lobe type is homozygous.

c Blood group is heterozygous.

d Ear lobe type is heterozygous.

e Blood group is homozygous but ear lobe type is heterozygous.

f Blood group is heterozygous but ear lobe type is homozygous.

g Blood group and ear lobe type are homozygous.

h Blood group and ear lobe type are heterozygous.

8 The following diagram represents part of a family tree showing the inheritance of tongue rolling. The allele D represents roller and d represents non-roller.

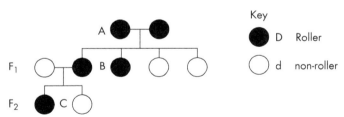

a Complete the table below for individuals A and C.

Individual	Possible genotype(s)	Phenotype
A		Roller
B	DD or Dd	Roller
C	dd	

b Individual C married a heterozygous roller and had four children.

Calculate the chance of any individual child being a roller.

c The same two individuals had four non-roller children.

Explain why the predicted ratio of roller children may be different to the actual ratio.

9 A type of blindness in humans is caused by a single gene.

The diagram below shows the pattern of inheritance in one family tree.

H represents the sighted form of the gene.

h represents the non-sighted form of the gene.

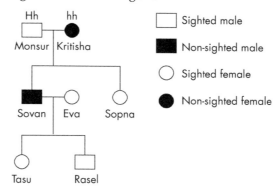

□ Sighted male

■ Non-sighted male

○ Sighted female

● Non-sighted female

a Using Monsur as an example, explain how it is known that the sighted form of the gene is dominant.

> **Hint** You must refer to both genotype and phenotype to answer this one correctly.

b Use information in the family tree to complete the following table to show the genotype and phenotype of each individual.

Individual	Possible genotype(s)	Phenotype
Sovan		
Rasel		

c Sopna marries a man who has the same genotype as her. State the chance of their child being able to see.

Exercise 10C Key terms

Link each term below with the correct description.

	Term		Description
1	variation	**A**	predicted ratio of phenotypes or genotypes in a genetic cross
2	discrete variation	**B**	second generation of offspring usually from crossing the first generation
3	continuous variation	**C**	actual ratio produced after a genetic cross
4	single gene inheritance	**D**	parental generation in a genetic cross
5	polygenic inheritance	**E**	one gene controls one inherited trait. Shows discrete variation
6	phenotype	**F**	many genes control the inherited trait. Usually shows continuous variation
7	genotype	**G**	first generation of offspring from a genetic cross
8	allele	**H**	different form of a gene
9	dominant	**I**	characteristics which show a range of values between a minimum and a maximum
10	true breeding	**J**	characteristics which fall into distinct groups
11	recessive	**K**	both alleles are the same
12	homozygous	**L**	any difference between individuals of the same species
13	heterozygous	**M**	an organism's physical appearance, or visible traits
14	P	**N**	an organism's genetic makeup
15	F_1	**O**	an allele that is masked when a dominant allele is present. Need two to be expressed
16	F_2	**P**	an allele that is always expressed when present. Only require one to be expressed
17	monohybrid cross	**Q**	alternative term for homozygous
18	predicted ratio	**R**	a segment of DNA on a chromosome that codes for a specific protein
19	actual ratio	**S**	a cross in which only one characteristic is tracked
20	Punnett square	**T**	a chart that shows all the possible combinations of alleles that can result from a genetic cross
21	gene	**U**	two different alleles

Exercise 11A Structure and functions

1 Name three organs found in a plant.

2 Below is a diagram of the structure of a leaf.

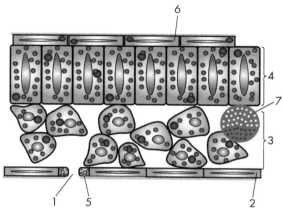

Complete the table below. Add the number from the diagram that matches with each structure.

Leaf structure	Number in diagram	Function/Description
Upper epidermis		
Palisade mesophyll		
Spongy mesophyll		
Vein		
Lower epidermis		
Guard cell		
Stoma		

3 Suggest two processes that occur within the roots that result in the absorption of water and minerals.

4 Describe how the structure of a root hair cell is related to its function.

5 State how water and minerals are transported from the roots up through the stem.

6 Complete the table below, stating the structure and function of the xylem and phloem vessels in a plant.

Vessel	Structure	Function
Xylem		
Phloem		

7 Describe the pathway of water through the plant, from the soil to the air.

> **Hint** This can be asked as a three-mark extended response question, so be sure to write an extended answer.

8 Describe the process of transpiration.

9 State how the following factors affect transpiration rate.

a Increase in environmental temperature.

b Decrease in surface area of leaf.

c Increase in environmental humidity.

d Decrease in environmental wind speed.

10 Which leaf structure contains air spaces for more efficient gas exchange?

A Palisade mesophyll   B Spongy mesophyll

C Stomata   D Upper epidermis

11 Which leaf structure is thin and transparent to allow maximum light transmission?

A Palisade mesophyll   B Spongy mesophyll

C Stomata   D Upper epidermis

12 Which of the following changes would cause an increase in transpiration rate?

A Increase in environmental humidity   B Decrease in environmental temperature

C Decrease in environmental wind speed   D Increase in surface area of leaf

13 Decide if each of the following statements is **TRUE** or **FALSE** and tick the correct box.

If the answer is **FALSE**, write the correct word(s) in the correction box to replace the word underlined in the statement.

Statement	True	False	Correction
Three examples of plant organs are roots, stems and <u>seeds</u>.			
Water and minerals enter the plant through the root hairs and are transported in dead <u>xylem</u> vessels.			
The <u>palisade</u> mesophyll has many chloroplasts for efficient photosynthesis.			
Transpiration rate <u>increases</u> when humidity increases.			
<u>Water</u> is transported up and down the plant in living phloem.			

1 A group of students set up an experiment to test the effect of different temperatures on the rate of transpiration in a plant. A potometer can be used to measure transpiration rate. They set up potometers as shown below.

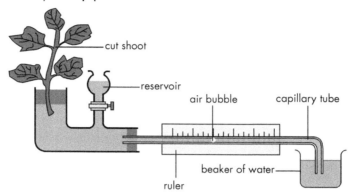

They placed potometers and plants into sealed containers at seven different temperatures and noted the distance travelled by the air bubble along the ruler in one minute. The distance travelled per minute is a measure of transpiration rate.

The results of their experiment are shown in the table below.

Temperature (°C)	Distance travelled along the ruler (mm) in one minute
0	0
10	2
20	9
30	10
40	5
50	2
60	0

a Describe the relationship between temperature and transpiration rate in this experiment.

b Give a conclusion that the students could make based on these results.

Hint   Your conclusion should relate to the aim.

c State one variable kept constant during this experiment.

Hint   Variables must be kept constant to increase validity.

d State one variable, not mentioned above, that would also have to be kept constant.

e Describe one way in which the reliability of the results could be increased.

f Describe a control experiment that could be set up alongside this experiment.

g Describe the purpose of a control experiment.

h Calculate the percentage increase in distance travelled by the air bubble from 10°C to 20°C.

i State the simple whole number ratio of the distance travelled by the air bubble at 10°C compared with distance travelled at 30°C.

j Suggest a reason for the result at 60°C.

Exercise 11C Key terms

Link each term below with the correct description.

1 upper epidermis

A collective name applied to the xylem and phloem together

2 palisade mesophyll

B the single layer of protective cells on the upper surface of a leaf

3 spongy mesophyll

C the loss of water by evaporation through the leaves

4 lower epidermis

D small openings found mostly on the underside of a leaf through which oxygen and carbon dioxide can diffuse and water can exit via evaporation

5 guard cells

E regulates the activity of phloem such as the uptake of sugar to provide energy

6 stomata

F protective layer of cells on the bottom of a leaf which contains stomata and guard cells

7 root hair

G process where water is lost through the stomata in the leaves

8 xylem

H pores through which sugar is transported from cell to cell

9 phloem

I a pair of cells that surround stomata and control their opening and closing

10 lignin

J loose tissue beneath the palisade layer of a leaf; has many air spaces between its cells for diffusion of gases

11 transpiration

K living cells which transport sugar in both directions in a plant

12 evaporation

L layer of tall cells with many chloroplasts and is the site where most photosynthesis takes place

13 companion cell

M found in xylem vessels for support; to withstand pressure changes as water moves through plant

14 sieve plate

N dead cells forming tubes which transport water and minerals up the plant

15 vein

O absorbs the nutrients and water from the soil; has large surface area to increase absorption

12 Transport systems – animals

Exercise 12A Blood

1 State the three components of mammalian blood.

2 State three substances that mammalian blood transports.

3 State the function of a red blood cell.

4 Describe three structural adaptations of a red blood cell.

5 Explain how the structure of a red blood cell relates to its function.

> **Hint** This can be asked as a three-mark extended response question, so be sure to write an extended answer.

6 Write the word equation for the reversible reaction involving oxygen and haemoglobin.

7 State the system to which white blood cells belong.

8 State one function of white blood cells.

9 **a** State the two main types of white blood cells.

   b Describe how each type carries out its function.

10 Describe the relationship between antibody and pathogen.

Exercise 12B Heart and circulatory system

1 The diagram below shows the human heart.
Identify the labelled structures.

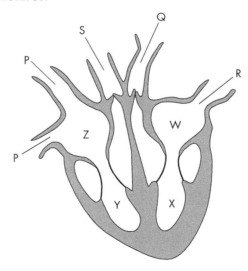

2 State which letters in the diagram of the human heart below show

   a Valves

   b Chambers

   c Blood vessels

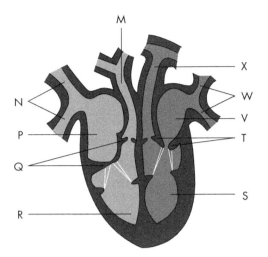

3 Complete the diagram below which shows the flow of blood in the body. Colour the circles with oxygenated blood red and the circles with deoxygenated blood blue.

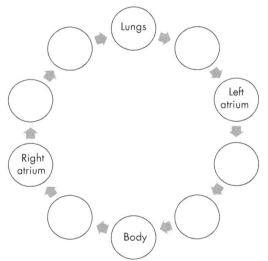

4 Complete the table below.

Heart structure	Function
Atrium	
Ventricle	
Heart valve	
Aorta	
Pulmonary artery	
Pulmonary vein	
Vena cava	
Coronary artery	

5 Complete the table below.

Blood vessel	Structure	Function
Artery		
Vein		
Capillary		

6 Name the type of blood vessel that:

a contains valves

b transports blood away from the heart

c transports blood into the heart

d allows efficient exchange of materials

e has a wide central channel

f has a narrow central channel

g has thick walls

h has thin walls

i has walls one cell thick

j transports blood at high pressure

k transports blood at low pressure

l is where most diffusion of materials occurs between the blood and cells

7 State the name of the blood vessel that supplies the heart muscle with blood.

8 State the function of valves in the circulatory system.

9 Compare any two of the following: arteries, veins or capillaries.

> **Hint** Another command word: 'compare' means give the key features of two different items or ideas and discuss or state their similarities.

> **Hint** This can be asked as a three-mark extended response question, so be sure to write an extended answer.

10 Which of the following describes the structure and function of a vein?

	Structure	Function
A	Thin walls with valves	Transports blood into the heart at low pressure
B	Narrow central channel	Transports blood away from the heart at low pressure
C	Thicker walls with valves	Transports blood away from the heart at high pressure
D	Wide central channel	Transports blood into the heart at high pressure

11 Which of the following structures in the heart is the first to receive deoxygenated blood from the body?

A Right atrium   **B** Left atrium

C Right ventricle   **D** Left ventricle

12 Which of the following vessels takes oxygenated blood to the body?

A Vena cava   **B** Aorta

C Pulmonary artery   **D** Pulmonary vein

13 Which of the following vessels takes deoxygenated blood to the lungs?

A Vena cava   **B** Aorta

C Pulmonary artery   **D** Pulmonary vein

14 Decide if each of the following statements is **TRUE** or **FALSE** and tick the correct box.

If the answer is **FALSE**, write the correct word(s) in the correction box to replace the word underlined in the statement.

Statement	True	False	Correction
In mammals the blood contains plasma, red blood cells and white blood cells.			
Red blood cells are specialised by being circular in shape, having no nucleus and containing haemoglobin.			
Phagocytes carry out phagocytosis by engulfing pathogens.			
Arteries have thick, muscular walls, a wide central channel, and carry blood under high pressure away from the heart.			
Capillaries are thin-walled and have a small surface area, forming networks at tissues and organs to allow efficient exchange of materials.			

Exercise 12C Key terms

Link each term below with the correct description.

1. nutrients, oxygen, carbon dioxide
2. oxygenated
3. deoxygenated
4. blood
5. heart
6. engulf
7. atrium
8. ventricle
9. aorta
10. pulmonary vein
11. pulmonary artery
12. vena cava
13. artery
14. vein
15. capillaries
16. coronary artery
17. valve
18. red blood cells
19. haemoglobin
20. oxyhaemoglobin
21. white blood cell
22. phagocyte
23. lymphocyte
24. pathogens
25. antibody

A upper, smaller chambers of the heart; blood enters in veins

B transported in the blood in mammals

C transport oxygen; are specialised by being biconcave in shape, having no nucleus and containing haemoglobin

D protein that carries oxygen in red blood cells

E takes deoxygenated blood to the lungs from the right ventricle

F cell found in blood, involved in defence of the body against pathogens

G type of white blood cell which undergoes phagocytosis to engulf pathogens

H supplies heart tissue with blood

I substance containing plasma, white blood cells and red blood cells; transports nutrients, oxygen and carbon dioxide

J returns oxygenated blood from lungs into the left atrium

K returns deoxygenated blood from the body into the right atrium

L pumps blood to lungs and body

M phagocytes do this to pathogens

N Y-shaped molecule secreted by lymphocytes to bind to a specific pathogen

O lower, larger chambers of the heart; blood exits in arteries; left is thicker than right

P largest artery, exiting the left ventricle

Q have thick, muscular walls, a narrow central channel, and carry blood under high pressure away from the heart

R haemoglobin combined with oxygen

S foreign cells within the body such as bacteria, fungi or viruses

T found in veins to prevent backflow of blood

U form networks at organs and tissues, are thin-walled and have a large surface area, allowing exchange of materials

V carry blood under low pressure back to the heart; have thinner walls and a wide channel

W rich in oxygen

X low in oxygen

Y type of white blood cell which secretes antibodies which bind to a specific pathogen

13 Absorption of materials

Exercise 13A Features and functions

1
  a Give two examples of substances that must be absorbed into the bloodstream to be delivered to cells.

  b State the ATP-producing process that requires these substances.

  c State what must be removed from cells during this process.

  d Name the process that removes this substance.

  e State the fate of this substance after leaving the cell.

2
  a Explain how tissues are able to exchange materials so efficiently at the cellular level.

  b State the three features that surfaces involved in absorption of these materials all have in common.

  c Explain how these features impact upon absorption.

3 Name the gas-exchange organs in the human body.

4 Describe the structure of these organs.

5 Explain how the structure increases efficiency of absorption.

> **Hint** This can be asked as a three-mark extended response question, so be sure to write an extended answer.

6 State the two gases absorbed through the walls of this organ.

7 Assign each word from the list below to a letter in the diagram

capillary/alveolus/oxygen/carbon dioxide

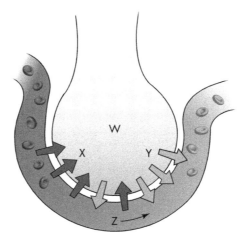

8 State the organ where nutrient absorption occurs.

9
  a Describe how nutrients from food are absorbed into the blood.

  b Give the name of the structure where this absorption occurs.

10 Describe how the structure of a villus allows it to absorb nutrients efficiently.

11 Which of the following structures absorbs glucose from digested food into the blood?

A Lacteal   B Alveolus

C Villus   D Artery

12 Which of the following structures absorbs glycerol from digested food?

A Lacteal   B Alveolus

C Villus   D Artery

13 Which of the following structures allows gas exchange between the lungs and the blood?

A Lacteal   B Alveolus

C Villus   D Artery

14 Identify the structures labelled in the diagram below of a villus.

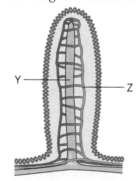

15 Decide if each of the following statements is **TRUE** or **FALSE** and tick the correct box.

If the answer is **FALSE**, write the correct word(s) in the correction box to replace the word <u>underlined</u> in the statement.

Statement	True	False	Correction
Nutrients from food and <u>carbon dioxide</u> must be absorbed into the bloodstream to be delivered to cells for respiration.			
Tissues contain <u>arterial</u> networks to allow the exchange of materials at cellular level.			
Surfaces involved in the absorption of materials have certain features in common: large surface area, <u>thin</u> walls, extensive blood supply.			
Oxygen and carbon dioxide are exchanged across the thin <u>alveolar</u> walls to or from the many blood capillaries which surround each alveolus.			
Each villus contains a network of capillaries to absorb glucose and <u>proteins</u> and a lacteal to absorb fatty acids and glycerol.			

Exercise 13B Key terms

Link each term below with the correct description.

1 capillary networks

2 features of surfaces used for exchange

3 alveolus

4 villus

5 lacteal

6 small intestine

7 respiration

8 diffusion

9 carbon dioxide

10 glucose

11 oxygen

A found within the villus, where fatty acids and glycerol are absorbed

B organ of the digestive system where many villi are found

C large surface area, thin walls, extensive blood supply

D process where glucose is broken down to produce ATP

E small, sac-like structures where diffusion of gases takes place between the lungs and blood

F process where molecules move from an area of high concentration to an area of low concentration

G waste material produced by cells during respiration

H respiratory substrate

I allow the exchange of materials at cellular level between the cells and tissues

J substance that diffuses from alveoli into blood and is essential for aerobic respiration

K small, finger-like projections where the products of digestion pass from the small intestine into the blood or lacteal

14 Ecosystems

Exercise 14A Ecosystems

1 Using the word bank provided, describe what is meant by the term biodiversity. Complete the sentences by writing in the appropriate word(s).

A word may be used once, more than once, or not at all.

plants – habitat – living – non-living – interact – ecosystem

An _____consists of all the animals and _____ living in a particular _____ as well as the _____ components with which the organisms _____.

2 A student was examining an area of woodland and found a number of small animals that included snails, earthworms, slugs, beetles and woodlice amongst small stones and tree bark.

a Which of the following best describes the animals, stones and tree bark?

    A Ecosystem   **B** Habitat

    C Community   **D** Population

b Which of the following best describes all of the earthworms?

    A Ecosystem   **B** Habitat

    C Community   **D** Population

3 An organism that can synthesise its own food by photosynthesis is called a

    A producer   **B** consumer

    C herbivore   **D** species

4 A leopard is an animal that hunts and eats other animals.

Which of the following terms best describes the leopard?

    A Omnivore   **B** Carnivore

    C Herbivore   **D** Consumer

5 The total variety of all organisms on Earth is termed

    A habitat   **B** biodiversity

    C ecosystem   **D** population

6 Look at the following simple food chain:

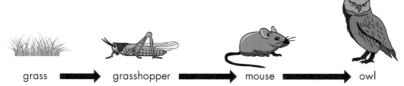

grass ➡ grasshopper ➡ mouse ➡ owl

a State the organism that is the producer.

b Name one organism that is a prey animal.

c Explain what the arrows indicate.

> **Hint**   Don't write 'shows what eats what', but rather consider energy flow and its direction.

7 A pond community could consist of all of the

A water boatmen   B animals and plants

C water fleas   D plants

8   a   Describe what is meant by the term biodiversity.

    b   Suggest three examples of human activity that could have a negative effect on biodiversity.

9 The diagram below shows a simple food web.

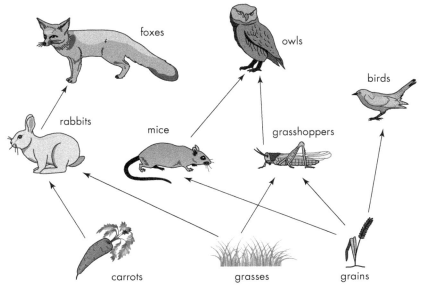

a   Complete the food web by drawing two arrow-headed lines to show that the birds are both prey and predator.

b   State the ultimate energy source to support this food web.

c   Name two producers shown in this food web.

d   Name two consumers shown in this food web.

e   Predict the effect in the short term on the population of grasshoppers if the mice were removed from this food web. Explain your answer.

f   Write out a food chain consisting of four organisms from this food web.

10 The following table shows some feeding relationships.

Animal	Feeding relationships
Kingfisher	eats small fish and frog
Small fish	eats tadpole
Tadpole	eats algae
Frog	eats water beetle and snail
Water Beetle	eats algae
Snail	eats algae

Construct a possible food web for these feeding relationships.

11 The table below shows some features of food chains and food webs.

Decide if each statement is **TRUE** or **FALSE** and tick the correct box.

If the statement is **FALSE**, write the correct word(s) into the correction box to replace the word(s) underlined in the statement.

Statement	True	False	Correction
A food chain is a diagram that shows where one organism feeds on the previous organism.			
In a food web an animal that eats other animals is known as a producer.			
The arrow-headed lines in a food chain show the direction of energy flow.			

Exercise 14B Niche

1 In a freshwater pond, a water flea is a prey animal to tadpoles, dragonflies and sticklebacks.

Which of the following best describes the niche of the water flea?

A The pond where it lives   B Its role in the pond

C Food supply to its prey animals   D The plants and animals in the pond

2 Using the word bank provided, complete the sentences by writing in the appropriate words.

A word may be used once, more than once, or not at all.

nutrients – resources – light – competition – community – role – suitable temperature – ecosystem – interacts – predation

A niche is the _____ an animal or plant plays within a _____. It is a function of the _____ required by an organism from its _____. These include _____, and availability of _____ as well as how an organism _____ with other organisms. A niche involves both _____ and _____ as well as the particular conditions it can survive in such as _____ _____.

Exercise 14C Competition

1 Explain why intraspecific competition is more intense than interspecific competition.

2 In which of the following would competition not take place?

A Hawks and owls hunting for mice   B Water plants and small invertebrates in a pond

C Different weeds growing on a lawn   D Cows feeding in a field

3 Which of the following describes intraspecific competition?

	Species involved	Resources required
A	different	one or few
B	different	all
C	same	one or few
D	same	all

4 Which of the following describes interspecific competition?

	Species involved	Resources required
A	different	one or few
B	different	all
C	same	one or few
D	same	all

Hint Think of an 'intercity train' that travels from two or more different cities. Interspecific competition is between <u>different</u> species.

5 Some different species of birds are shown below along with the type of food they eat.

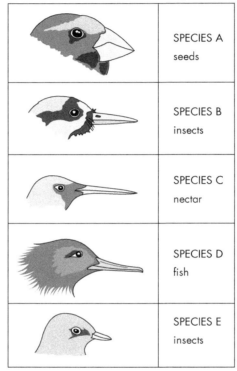

	SPECIES A seeds
	SPECIES B insects
	SPECIES C nectar
	SPECIES D fish
	SPECIES E insects

a State the feature shown that allows some birds to have different diets.

b Predict two species of bird that would show interspecific competition if the insect population declined.

c Give the term that describes the role of any one of these species of birds within its community.

Exercise 14D Key terms

Link each term below with the correct description.

1. species   **A** organism that can make its own food by photosynthesis

2. biodiversity   **B** group of interbreeding organisms whose offspring are fertile

3. population   **C** range of different species

4. producer   **D** organism that eats another organism as a source of energy

5. consumer   **E** organism that feeds only on plant material

6. herbivore   **F** animal that hunts other animals as its energy source

7. carnivore   **G** animal that is hunted by other animals

8. omnivore   **H** linear feeding relationship of organisms

9. predator   **I** a number of food chains linked together

10. prey   **J** organism that eats both plants and animals

11. food chain   **K** animal that feeds only on other animals

12. food web   **L** group of organisms all belonging to the same species

13. niche   **M** between individuals of the same species

14. intraspecific   **N** between individuals of different species

15. interspecific   **O** living things and their non-living environment

16. ecosystem   **P** role an organism plays within an ecosystem

17. habitat   **Q** general term for place in an environment where an organism lives

18. community   **R** collection of animals and plants living together in the same habitat

> **Hint** A 'mind map' could be a good way of remembering all the different terms used in this unit.

Exercise 14E Skills of scientific inquiry

1. A student collected data about the relative abundance (%) of five different invertebrates found in a pond water sample.

   The table below shows the results.

Invertebrate	Relative abundance (%)
Water flea	40
Freshwater shrimp	15
Snail	10
Water boatmen	25
Worm	10

a Present this data as a pie chart by completing the diagram below.

> **Hint** You need to be able to present data in a variety of different formats such as a bar graph, line graph, pie chart or a table.

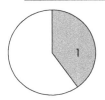 ▢ 1 – water fleas

b In total, the student counted 200 invertebrates.

State how many of these would have been water fleas.

2 The table below shows the growth of four different trees in different conditions.

Tree	Acid soil	Temperature		High light intensity
		Low	High	
1	✓	X	✓	✓
2	X	✓	X	X
3	✓	✓	X	✓
4	X	X	✓	X

X = poor growth   ✓ = good growth

State which tree you would expect to grow best in

a cool bright conditions where the soil is acidic

b acidic soil in bright warm conditions.

3 The two graphs below relate to a small herbivore that breeds throughout the year.

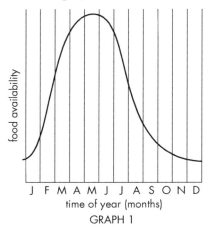
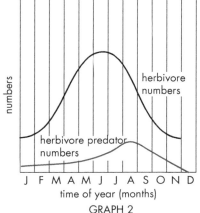

Graph 1 shows the changes in the food availability over the year where the herbivore lives.

Graph 2 shows the changes in the numbers of the herbivores and one of its predators.

The following are possible conclusions from this data.

1 Numbers of the herbivores rise in spring due to an increase in food availability.

2 The herbivores leave their preferred habitat in late summer due to a lack of food availability.

3 Predator numbers decrease in autumn due to a decrease in herbivore numbers.

State which of these conclusions is (are) correct based on the data available.

15 Distribution of organisms

Exercise 15A Competition for resources

1 Using the word bank provided, complete the sentences by writing in the appropriate word(s).

A word may be used once, more than once, or not at all.

biodiversity – species – ecosystem – grazing

_____ is carried out by field herbivores such as sheep and cows that feed on plant materials.

At low _____ intensities, the _____ of a field is low because a few _____ of plants are able to successfully compete against the others and dominate the _____.

2 The following factors affect living things in their habitat.

pH – light intensity – moisture level – temperature – nutrient availability – predators – competition – disease

Construct a table to divide these factors into those that are biotic and those that are abiotic.

Exercise 15B Measuring abiotic factors

1 A group of students studied a freshwater stream where they measured some abiotic factors that included pH and temperature of the water in the early morning.

  a The students used a thermometer to measure the water temperature.

    Describe how they could have improved on the accuracy of the temperature readings.

  b State one other abiotic factor that they could have included in their measurements.

  c State one design flaw in this investigation and how it could be improved on.

2 One possible source of error using a pH probe may be due to contamination by soil left on the probe from a previous reading.

Describe how this error could be reduced.

3 When using a light meter, which of the following will both help to reduce errors?

  A Sampling at the same time of day and ensuring the weather conditions are constant

  B Holding the meter and making sure there is no rain falling

  C Taking readings immediately and ensuring the Sun is shining

  D Looking closely at the meter and taking many readings

Exercise 15C Sampling

1 Which of the following is a possible source of error when using a quadrat to sample a lawn for the abundance of buttercup?

  A Using a large number of samples **B** Placing the quadrats in interesting areas

  C Always using the same quadrats **D** Throwing quadrats randomly

2 The diagram shows a particular piece of equipment used in sampling.

a Name this piece of equipment.

b Predict the kind of animal liable to be captured by this equipment.

c State one reason why the stone cover is set in place.

d State one possible source of error using this piece of equipment and explain how this can be reduced.

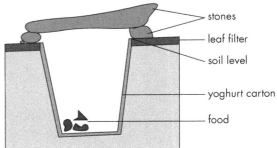

stones
leaf filter
soil level
yoghurt carton
food

3 A student used a 0.25 m by 0.25 m quadrat and took five random samples to estimate the abundance of buttercups in a large lawn.

The results of the five samples are shown below.

Sample	Abundance of buttercups
1	30
2	25
3	40
4	0
5	50

a State which sample is markedly different from the other four.

Suggest a possible reason for this result.

b State two ways the student could improve on the design of this experiment.

Exercise 15D Keys

1 Some features of five wild flowers are recorded in the table below.

Flower	Leaf shape	Colour of petals	Number of petals
Bluebell	Narrow	Blue-coloured and bell-shaped	6
Wild daffodil	Narrow	Yellow-coloured and trumpet-shaped	6
Red nettle	Broad	Reddish purple-coloured	2
Lesser celandine	Broad	Orange-coloured and daisy-shaped	Variable
Wild primrose	Broad	Pale yellow-coloured	4

a Construct a paired-statement key for these flowers using the information in the table.

b Describe the lesser celandine.

2 The diagram below illustrates five different invertebrates.
Complete the key using only features visible on the diagram.

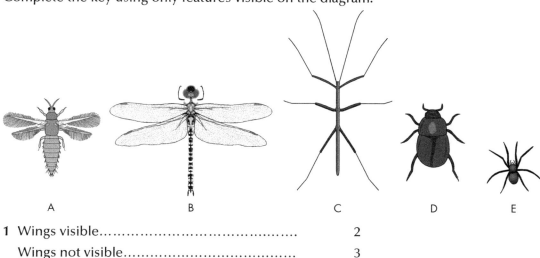

A   B   C   D   E

1 Wings visible...................................... 2
   Wings not visible................................. 3

> **Hint** There is usually more than one possible answer to this type of question.

3 The following key refers to evergreen trees.

> **Hint** Don't be put off by these formal tree names. Work through the key systematically.

1 Leaves scaly, small blue cones   2
   Leaves needle-shaped, large brown cones   3
2 Leaves rough, broad cones, forked trunk   *Juniperus osteosperma*
   Leaves smooth, narrow cones, trunk not forked   *Juniperus scopulorum*
3 Needles in bundles, cones have woody scales   4
   Needles not in bundles, cones have papery scales 5
4 Needles long, trunk straight and tall   *Pinus contorta*
   Needles short, trunk short and bushy   *Pinus remota*
5 Needles flat and blunt   6
   Needles square   *Picea pungens*
6 Needles point outward   *Pseudotsuga menziesii*
   Needles bend upward   *Abies concolor*

a *Pinus contorta* has

   A a straight trunk and cones with papery scales

   B needles in bundles with large brown cones

   C scaly leaves and cones with woody scales

   D small blue cones and a tall trunk.

b From the key, state four features of *Abies concolor*.

c Using features from the key, state how many trees have a trunk that is forked.

d State the two trees being referred to from this description:

   have flat needles, large cones that have papery scales

Exercise 15E Effect of biotic and abiotic factors

1 State two reasons why grazing by livestock is important in habitat management.

2 Using the word bank provided, complete the sentences by writing in the appropriate word(s).

A word may be used once, more than once, or not at all.

abiotic – warm – temperatures – cold – nutrients – sunlight

Biotic and _____ factors affect the survival and growth of an organism. For example, if there is little _____ then plants may die because they are unable to photosynthesise. Water has the ability to float when it is frozen so that many aquatic organisms can survive in extremely _____ conditions even though the surface is frozen. There is a wide range of _____ within which the water still remains liquid with the ability to dissolve many essential _____ and minerals.

3 Which of the following are both abiotic factors that can affect biodiversity?

A pH and the presence of predators   B Disease and lack of food

C Temperature and rainfall   D Pollution and invasion of new species

4 Which of the following are both biotic factors that can affect biodiversity?

A Space and absence of predators   B Disease and nutrient availability

C Temperature and water availability   D Pollution and invasion of new species

Exercise 15F Indicator species

1 The table below shows a statement associated with an indicator species.

Decide if the statement is **TRUE** or **FALSE** and tick the correct box.

If the statement is **FALSE**, write the correct words into the correction box to replace the words <u>underlined</u> in the statement.

Statement	True	False	Correction
Indicator species by their <u>presence only</u> can indicate the level of pollution in the environment.			

2 State the term used to describe organisms such as lichens that are very sensitive to levels of acid in rain.

Exercise 15G Key terms

Link each term below with the correct description.

1   grazing

2   biotic factor

3   abiotic factor

4   sample

5   quadrat

6   pitfall trap

7   paired-statement key

8   indicator species

9   pollution

10   source of error

A predetermined number of observations taken from a larger population

B device for sampling organisms in a predetermined area

C device for sampling invertebrates living on soil surface

D device for naming unknown organisms using a number of listed descriptions grouped in twos

E organism whose presence or absence is a function of the state of its habitat

F presence of potentially harmful substances in the environment

G method of feeding in which a herbivore feeds on plants

H variable that has a living origin and affects biodiversity

I flaw in design of an investigation that could give inaccurate or unreliable results

J variable which has a non-living origin and affects biodiversity

Exercise 15H Skills of scientific inquiry

1   Using a digital meter and probe, a student examined the pH of soil at two different depths, 1 cm from the surface and 10 cm from the surface.

She took five different samples within an area of 1 m².

The table below shows the results.

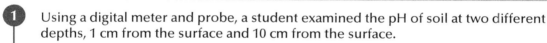

	pH readings			pH readings
	6.7			7.2
Probe inserted 1 cm from surface	7.0	Probe inserted 10 cm from surface		6.9
	7.2			7.2
	7.4			7.8
	6.7			6.9

a   Calculate the average pH reading for each insertion depth.

b   State the range of pH values for each depth.

> **Hint**   There are two ways of answering this. You can give the lowest pH to the highest pH or the difference between these values.

c   Suggest two design errors in the set up for this investigation and how these could be corrected.

d   State what type of factor this student was investigating.

2 A student carried out an investigation to estimate the number of a particular type of mushroom growing on part of a forest floor.

He used a quadrat measuring 20 cm by 20 cm and took ten random samples in four different areas.

The averaged results are shown in the table below.

Sample point	1	2	3	4
Average number of mushrooms	8	7	6	9

Which of the following changes to his design would have improved the reliability of the results?

A Sample only one area of the forest floor   B Use quadrats that were larger in size

C Record fewer than ten samples   D Include other varieties of mushrooms

3 A researcher used a quadrat 0.5 m by 0.5 m to estimate the abundance of three different plants, before and after treatment with a weedkiller, on an area of grassland that sloped.

The results are shown in the table below.

Plant	Average abundance before treatment	
	Top of slope	Base of slope
Grass	40	40
Dandelion	30	15
Common chickweed	20	25

Plant	Average abundance after treatment	
	Top of slope	Base of slope
Grass	40	40
Dandelion	10	3
Common chickweed	16	20

a Calculate the percentage of dandelions killed by the weedkiller at the base of the slope.

> **Hint** Make sure you can handle percentages. Remember it's the difference over the original × 100.

b Identify which plant was not affected by the weedkiller.

c Identify which plant, before treatment, showed the most marked difference in abundance between the top and base of the slope.

d Name one weed and state its slope position that showed a 20 % reduction in abundance after treatment.

e In order to make the results reliable, describe how the researcher should have used the quadrat.

f State one abiotic factor that could have affected the distribution of the plants from the top of the slope to the base of the slope.

16 Photosynthesis

Exercise 16A Two-stage process

1 Using the word bank provided, complete the sentences by writing in the appropriate word(s).

A word may be used once, more than once, or not at all.

two – light – food – photosynthesis – chlorophyll – hydrogen – chloroplasts – oxygen – ATP – water – glucose (sugar) – carbon

_____ is a _____ stage process that occurs in green plants allowing them to make their own _____ . In the first stage, _____ energy from the Sun is trapped by the green pigment _____ in the organelles called _____. This allows for the synthesis of the energy-rich molecule _____. In this stage _____ is split into oxygen and _____.

In the second stage, _____ is fixed to form the carbohydrate _____ while the gas _____ diffuses from the cell.

2 Which of the following is a raw material used in photosynthesis?

A Starch   **B** Cellulose

C Carbon dioxide   **D** Glucose

3 The statements below refer to various processes associated with photosynthesis.

Decide if each statement is **TRUE** or **FALSE** and tick the correct box.

If the statement is **FALSE**, write the correct word(s) into the correction box to replace the word(s) <u>underlined</u> in the statement.

Statement	True	False	Correction
Green plants use <u>ATP</u> as their primary source of energy.			
The light reactions take place in the <u>cytoplasm</u> of a green plant cell.			
<u>Oxygen</u> gas diffuses out of a plant cell during the light reactions.			

4 **a** State the biological molecules that control the rate of the light reactions.

b State one abiotic factor that would affect the rate of activity of these biological molecules.

5 The diagram below shows some of the processes involved in photosynthesis.

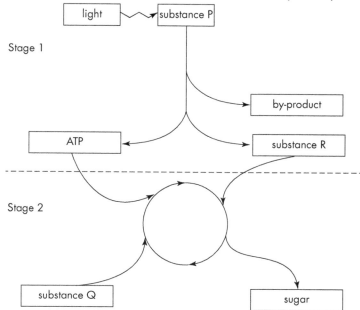

a Identify substances P, Q and R.

b Name Stage 2.

c State the 'by-product' produced in Stage 1.

d State three possible uses of the sugar produced in Stage 2.

e Describe the role of carbon dioxide in stage 2.

6 The diagram opposite shows a 'half-leaf' experiment investigating the factors that affect photosynthesis in a green plant.

a After several hours of bright illumination, predict the result of testing leaf X for the presence of starch.

b Explain what this result shows regarding the factors required for photosynthesis to take place.

c Describe how the experiment could be modified to show that light is needed for photosynthesis to take place.

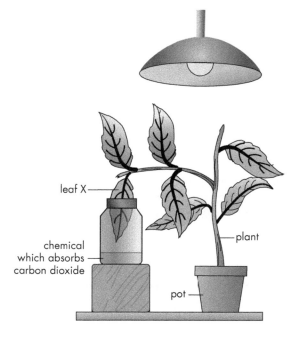

7 **a** Explain why photosynthesis is affected by temperature.

> **Hint** Try to link different parts of the course here. Think, for example, what controls the stages in photosynthesis.

b Explain why a temperature in excess of 50°C would prevent photosynthesis from taking place.

8 In the first stage of photosynthesis

A oxygen is required **B** light is not required

C water is split **D** glucose is produced

9 In the carbon fixation stage of photosynthesis

A carbon dioxide is produced **B** ATP is used up

C hydrogen is produced **D** light is required

10 The table below compares photosynthesis and aerobic respiration.

Insert the following into the appropriate spaces. Note that a term may be used more than once and more than one term might be used in a single box.

carbon dioxide – water – glucose – oxygen – sunlight – ATP – synthesised

Process	Raw materials	Products	Energy source
Photosynthesis			
Aerobic respiration			

11 During photosynthesis, the correct sequence for the synthesis of cellulose is

A sugar⋏starch⋏cellulose **B** sugar⋏carbon dioxide⋏cellulose

C starch⋏sugar⋏cellulose **D** carbon dioxide⋏sugar⋏cellulose

Exercise 16B Use of sugar synthesised

1 Which of the following is a storage carbohydrate in plants?

A Starch **B** Cellulose

C Carbon dioxide **D** Glucose

2 Which of the following is a potential substrate to be used in respiration by plants?

A Starch **B** Cellulose

C Carbon dioxide **D** Glucose

> **Hint** Remember that plants respire at a relatively constant rate to supply energy for growing, moving and transport, etc.

Exercise 16C Limiting factors

1 The graph below shows the effect of increasing light intensity on the rates of photosynthesis at two different temperatures.

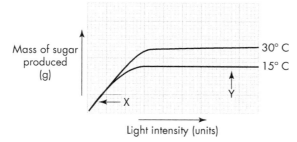

a Name the limiting factor at the two points indicated on the graph, X and Y.

b Explain why the rate of photosynthesis is greater at 30°C than at 15°C.

c State one factor, other than light intensity or temperature, that could affect the rate of photosynthesis.

d State how the rate of photosynthesis is measured here.

2 Which of the following factors would be mostly likely to limit the rate of photosynthesis in green plants in low light intensities?

A Carbon dioxide concentration   **B** Light intensity

C Temperature   **D** Oxygen availability

3 A limiting factor in photosynthesis

A does not stop an increase in the rate of photosynthesis

B eventually stops the rate of photosynthesis increasing

C reduces the rate of photosynthesis when in short supply

D will have no effect on the rate of photosynthesis

4 Which of the following is BOTH a potential limiting factor and a raw material for photosynthesis?

A Oxygen   **B** Carbon dioxide

C Hydrogen   **D** Glucose

5 The effect of temperature on the rate of photosynthesis for two different species of plants, A and B, is shown on the graph below.

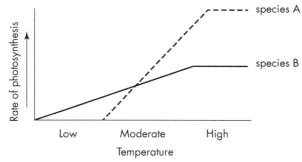

Which of the following conclusions is correct? The rate of photosynthesis of species A is

A higher than B at low temperatures   **B** lower than B at low temperatures

C lower than B at high temperatures   **D** lower than B at moderate temperatures

Exercise 16D Key terms

Link each term below with the correct description.

1. photosynthesis — A sugar formed during second series of reactions in photosynthesis

2. cellulose — B major component of plant cell walls formed from glucose molecules

3. light reactions — C green pigment found in plant cells

4. carbon fixation — D enzyme-controlled reactions that form the second stage of photosynthesis

5. chlorophyll — E energy-rich molecule formed during light reactions of photosynthesis

6. chloroplast — F by-product of the breakdown of water during the light reactions of photosynthesis used in carbon fixation

7. ATP — G by-product of the breakdown of water during the light reactions of photosynthesis, which diffuses out of the plant cell

8. hydrogen — H raw material needed in the light reactions that is split to release hydrogen and oxygen

9. oxygen — I storage carbohydrate formed from glucose molecules

10. water — J enzyme-controlled processes forming first stage of photosynthesis

11. glucose — K organelle where photosynthesis takes place in green plant cells

12. starch — L process by which green plants make their own food

13. limiting factor — M potential limiting factor when carbon dioxide concentration and temperature are both non-limiting in photosynthesis

14. light intensity — N potential limiting factor when carbon dioxide concentration and light intensity are both non-limiting in photosynthesis

15. carbon dioxide concentration — O variable which will slow down a reaction if in short supply or below optimum

16. temperature — P potential limiting factor when light intensity and temperature are both non-limiting in photosynthesis

Exercise 16E Skills of scientific inquiry

1 A student investigated the effect of light intensity on the rate of photosynthesis, measured by the number of bubbles of oxygen given off per minute, in a pondweed.

She used the apparatus shown, using the metre stick to measure the distance of the lamp from the pondweed as a way of altering the light intensity.

Her results are shown in the table below.

Distance between pondweed and lamp (m)	Number of bubbles of oxygen given off (per minute)
2.5	3
2.0	6
1.5	8
1.0	40
0.5	40

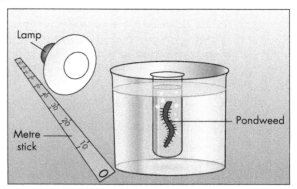

a Complete the following line graph to show this data.

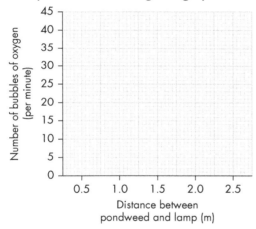

Mark the optimum rate of photosynthesis at the shortest distance between the lamp and the plant with an X on the graph.

b Describe the effect of increasing light intensity on the rate of photosynthesis.

c Suggest two ways the student could improve the design of the experiment.

d The following factors can affect the process of photosynthesis in green plants.

1 light intensity

2 carbon dioxide concentration in the water

3 temperature of the water

4 oxygen content of the water

Which two factors are both able to limit the rate of photosynthesis at light intensities greater than X?

A 1 and 2   B 1 and 4

C 2 and 3   D 3 and 4

17   Energy in ecosystems

Exercise 17A Food chains and energy

1 Describe what is shown by a food chain.

2 State three ways in which energy is lost at each level of a food chain. State one way in which a very small quantity of energy is used.

> **Hint**   This is commonly asked in examinations. Ensure you memorise the answer.

3 Look at the food web below and answer the following questions.

State the organism(s) that are

a producers

b consumers

c predators

d prey

e herbivores

f carnivores

g omnivores

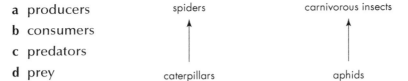

Exercise 17B Pyramids

1 Explain what is meant by:

a Pyramid of numbers.

b Pyramid of energy.

2 **a** For each of the following two food chains, draw a pyramid of numbers.

    i grass ⟶ zebra ⟶ lion ⟶ flea

    ii oak tree ⟶ caterpillar ⟶ blackbird ⟶ hawk

  b Draw a pyramid of energy for the oak tree food chain above.

  c Explain why the two pyramids have different shapes.

3 Decide if each of the following statements is **TRUE** or **FALSE** and tick the correct box.

If the answer is **FALSE**, write the correct word(s) in the correction box to replace the word <u>underlined</u> in the statement.

Statement	True	False	Correction
In transfers from one level to the next in a food chain, the majority of the energy is <u>gained</u> as heat, movement or undigested materials.			
Only a very small quantity of energy is used for <u>growth</u> and available at the next level in a food chain.			
A pyramid of <u>numbers</u> shows the energy present in each level of a food chain.			

4 In a food chain, which of the following would make energy available to the next stage?

A Heat production   B Movement

C Growth   D Elimination of undigested material

5 A food web is shown below.

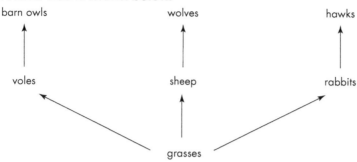

Which of the following describes the effect of removing sheep from this food web?

A Decrease in hawks   B Decrease in voles

C Increase in wolves   D Increase in rabbits

Exercise 17C Key terms

Link each term below with the correct description.

1 consumer   A heat, movement and undigested material

2 producer   B a diagram that shows the transfer of energy from one organism to the next

3 food chain   C a diagram to show the numbers of each organism at each level in a food chain

4 transfer of energy   D organism that eats both animals and plants

5 energy loss in a food chain   E organism that only eats animals

6 carnivore   F organism that only eats plants

7 herbivore   G a diagram to show the energy in organisms at each level in a food chain

8 omnivore   H represented by the arrow in a food chain

9 pyramid of energy   I an organism that gains energy from the Sun

10 pyramid of numbers   J an organism that gains energy by eating another organism

18 Food production

Exercise 18A Food yield, fertilisers and nitrates

1 State the current trend in human population and explain one effect of this trend.

2 **a** Give two examples of chemicals that would help increase food yield.

b Describe how the application of these chemicals would increase food yield.

3 Explain how nitrates enter plants.

4 State a function of nitrates in plants.

5 Describe how animals obtain nitrates.

6 State how nitrate levels in the soil can be artificially increased.

7 Describe how unwanted nitrates can enter fresh water.

8 Explain the effect of unwanted nitrates on fresh water.

> **Hint** This can be asked as a three-mark extended response question, so be sure to write an extended answer.

9 State one way in which fertiliser use can be reduced.

10 Which of the following correctly describes a result of excess fertilisers leaching into fresh water?

A Decrease in algal blooms   B Decrease in bacteria numbers

C Increase in oxygen levels   D Increase in nitrate levels

Exercise 18B Pesticides and alternatives

1 Describe one disadvantage of using pesticides.

2 Describe the effect of pesticides entering food chains.

3 State what is meant by bioaccumulation.

4 State two alternatives to the use of pesticides.

5 Describe what is meant by biological control.

6 Explain how the use of genetically modified (GM) crops might reduce the use of pesticides.

7 Decide if each of the following statements is **TRUE** or **FALSE** and tick the correct box.

If the answer is **FALSE**, write the correct word(s) in the correction box to replace the word underlined in the statement.

Statement	True	False	Correction
An increasing human population requires an <u>increased</u> food yield.			
Nitrates are used to produce amino acids, which are synthesised into plant <u>proteins</u>.			
Fertilisers can leach into fresh water, adding extra, unwanted <u>carbon</u>.			
Genetically modified (GM) crops can be used to reduce the use of <u>fertilisers</u>.			
As <u>pesticides</u> are passed along food chains, toxicity increases and can reach lethal levels.			

Exercise 18C Skills of scientific inquiry

1 The graph below shows the growth of the human population.

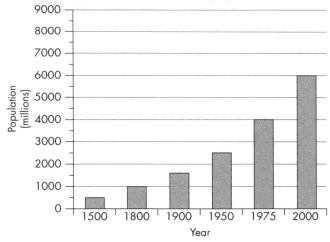

Use the data in the graph to answer the questions below.

a Describe the trend shown in the graph from 1950 onwards.

b Express the simple whole number ratio for the population in 1800 compared with 1950.

> **Hint** You should make the ratio as simple as possible by ensuring there are no more common denominators.

c Calculate the percentage increase in population from 1900 to 2000.

d Predict the human population in 2025, assuming the trend continues.

e Suggest one reason for the trend shown in the graph.

> **Hint** With predict questions, look at the last change in the graph/table and assume this change will occur again in the same time frame as the previous change.

Exercise 18D Key terms

Link each term below with the correct description.

1 food yield   **A** molecules that combine to make plant protein

2 fertiliser   **B** organism, usually a natural predator, introduced to eliminate pests

3 nitrate   **C** molecule made from amino acids

4 plant protein   **D** when fertilisers enter fresh water

5 pesticide   **E** explosion of growth of algae that cover the surface of the water, reducing sunlight available to the aquatic plants

6 GM crops   **F** when quantity of a chemical increases as it passes along a food chain

7 biological control   **G** total food produced

8 leaching   **H** the degree to which a substance can damage an organism

9 algal bloom   **I** chemical added to soil to increase growth/yield in plants

10 bioaccumulation   **J** genetically altered plants, often to increase resistance to pests or to increase growth/yield

11 toxicity   **K** used to produce amino acids in plants

12 amino acids   **L** chemical applied to plants to kill pests

19 Evolution of species

Exercise 19A Mutation

1. State what is meant by a mutation.

2. State three ways in which a mutation may affect an organism's survival chances.

3. Give the term used to describe mutations that are often unexpected or unpredictable.

4. Describe the importance of mutations in terms of the survival of a species.

5. State two environmental factors that can increase the rate of mutation.

Exercise 19B Adaptation and variation

1. State the importance of new alleles to plant and animal species.

   Hint   This is commonly asked in examinations. Ensure you memorise the answer.

2. State what is meant by the term adaptation.

3. Describe the importance of variation within a population.

Exercise 19C Natural selection

1. Describe one circumstance under which natural selection occurs.

2. Give another term for natural selection.

3. Describe the process of natural selection.

4. State one way in which natural selection occurs in species, with reference to offspring numbers.

Exercise 19D Speciation

1. Define speciation.

2. State what is meant by an isolation barrier.

3. Give three types of isolation barriers and give an example of each.

4. Give an account of the process of speciation.

   Hint   This can be asked as a three-mark extended response question, so be sure to write an extended answer.

5 Which of the following processes is the only source of new alleles?

A Mutation   B Natural selection

C Adaptation   D Evolution

6 Which of the following processes is described as survival of the fittest?

A Mutation   B Natural selection

C Adaptation   D Evolution

7 Which of the following processes is the change in an organism, over a long period of time, in response to changing environmental conditions?

A Mutation   B Natural selection

C Adaptation   D Evolution

8 Which of the following is an inherited characteristic that makes an organism well suited to survival in its environment/niche?

A Mutation   B Natural selection

C Adaptation   D Evolution

9 Decide if each of the following statements is **TRUE** or **FALSE** and tick the correct box.

If the answer is **FALSE**, write the correct word(s) in the correction box to replace the word <u>underlined</u> in the statement.

Statement	True	False	Correction
Mutations are spontaneous and are the only source of new <u>alleles</u>.			
New alleles produced by mutation can result in plants and animals becoming better <u>adapted</u> to their environment.			
Species often produce <u>fewer</u> offspring than the environment can sustain.			
Natural selection or survival of the fittest occurs when there are <u>variation</u> pressures.			
<u>Speciation</u> occurs after part of a population becomes separated by an isolation barrier.			

Exercise 19E Key terms

Link each term below with the correct description.

1 mutation

A a group of similar organisms that can breed and produce fertile offspring

2 environmental factors that increase the rate of mutation

B natural forces that promote the reproductive success of some individuals more than others

3 adaptation

C process resulting in the formation of a new species

4 importance of variation

D geographical, ecological or reproductive

5 species

E makes it possible for a population to evolve over time in response to changing environmental conditions

6 evolution

F the best adapted individuals survive to reproduce, passing on the favourable alleles that confer a selective advantage

7 natural selection (survival of the fittest)

G a random change to genetic material

8 selection pressures

H a change in a species over time

9 speciation

I radiation and chemicals

10 types of isolation barrier

J a trait that helps an organism survive and reproduce; can be structural or behavioural

Scientific literacy

Exercise 1

https://www.parkinson.org/blog/science-news/science-article/low-fat-dairy-parkinsons-risk#:~:text=Risk%20of%20Parkinson's-,Low%2DFat%20Dairy%20Foods%20Associated%20with%20Modest%20Increased%20Risk%20of,7%20online%20edition%20of%20Neurology.

Higher consumption of low-fat dairy products such as milk, butter, cheese and yoghurt may be linked to a modest increase in risk of Parkinson's disease, according to new research.

People who consumed three or more servings of low-fat dairy every day were slightly more likely to develop Parkinson's disease than those who consumed less than one serving a day.

The research is published in the scientific journal *Neurology*.

In the study, researchers at Harvard analysed approximately 25 years of data on 80,736 women and 48,610 men.

Participants completed health questionnaires every two years and diet questionnaires every four years.

During that time, 1036 people developed Parkinson's.

There was no link between full-fat dairy and risk of Parkinson's.

But those who consumed at least three servings of low-fat dairy a day had a 1 % chance of developing Parkinson's over the 25-year period, compared to 0.6 % in those who consumed less than one serving per day.

a Suggest the aim of the research described in the passage.

b A dependent variable is what scientists measure or observe as a result of the changes they make in their investigation.

Identify the dependent variable in this investigation.

c Complete a table, with suitable headings, to show the results of this trial.

d State one conclusion scientists could draw from this study.

e Give a reason why it could be suggested that the results of the investigation might be unreliable.

Exercise 2

https://www.diabetes.org.uk/research/research-round-up/research-spotlight/research-spotlight-low-calorie-liquid-diet

Can diet alone reverse Type 2 diabetes?

Research funded by Diabetes UK and carried out by a team from Newcastle University has discovered that Type 2 diabetes can be reversed by an extremely low-calorie diet alone.

In an early stage clinical trial of 11 people, all reversed their diabetes by drastically cutting their food intake to just 600 calories a day for two months. And three months later, seven remained free of diabetes.

Under close supervision of a medical team, the participants' diet consisted of liquid diet drinks and non-starchy vegetables. They were matched to a control group of people without diabetes and then monitored over eight weeks. Insulin production from their pancreas and fat content in the liver and pancreas were studied.

After just one week, the Newcastle University team found that their pre-breakfast blood glucose levels had returned to normal.

A special MRI scan revealed the fat level in the pancreas of each of the 11 people had returned from an elevated level (8 %) to a normal (6 %) level. In step with this, the pancreas regained the normal ability to make insulin and, as a result, blood glucose after meals steadily improved.

a Suggest the aim of the research described in the passage.

b Identify the independent variable in this investigation.

c Calculate the percentage change in pancreatic fat levels during this study.

d State one conclusion scientists could draw from this study.

e Give a reason why it could be suggested that the results of the investigation might be unreliable.

Exercise 3

Scientists believe that different temperatures will change the rate of growth of a plant species that has just been discovered in Scotland.

In a study, the heights of four plants were measured at the start and after four weeks. At the start all plants were 20 cm tall.

The plants were kept at four different temperatures: 10° C, 20° C, 30° C, and 40° C. The volume of water and concentration of nutrients added to the soil of the plants were kept the same.

After four weeks, the change in the height of each plant was recorded.

The plant at 10° C was 25 cm tall, 20° C was 32 cm tall, 30° C was 36 cm tall, and 40° C was 28 cm tall.

a Suggest the aim of the research described in the passage.

b Identify the independent variable in this investigation.

c Complete a table, with suitable headings, to show the temperatures and the change in the height of the plants.

d State one conclusion scientists could draw from this study.

e Give a reason why it could be suggested that the results of the investigation might be unreliable.

Exercise 4

Environmental protection analysis was carried out on water samples from three rivers. Scientists wanted to discover if the pH of different rivers would result in different levels of oxygen saturation.

The River Styx had the highest pH at 8.0. It also had the highest oxygen saturation at 94 % compared to River Rapid which had the lowest oxygen saturation at 65 %.

The Running River had the lowest value for suspended solids at 4.0 mg/l, with an oxygen saturation of 91.5 %.

River Rapid had a suspended solids value of 5.6 mg/l and the lowest pH at 7.7 compared to a value of 7.9 for the Running River. The highest value for suspended solids was recorded in the River Styx with a value of 6.0 mg/l.

a Suggest the aim of the research described in the passage.

b Identify one independent variable in this investigation.

c Complete a table, with suitable headings, to show the pH, the oxygen saturation and the suspended solids for each analysis site.

Analysis site			
River Styx			
River Rapid			
Running River			

d State one conclusion scientists could draw from this study.

e Give a reason why it could be suggested that the results of the investigation might be unreliable.

National 5
BIOLOGY
For SQA 2019 and beyond

**Mixed Exam
Question Practice**
Billy Dickson, Graham Moffat

MULTIPLE-CHOICE QUESTIONS

1 Red blood cells were placed into a salt solution more concentrated than blood plasma.

Which word best describes the predicted appearance of the cells after a few seconds in this solution?

A Burst

B Plasmolysed

C Turgid

D Shrunken

1

2 The diagram below shows molecules present in the cell membrane.

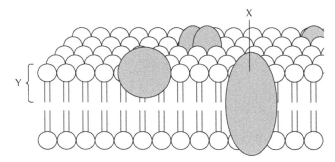

Which line in the table correctly identifies molecules X and Y?

	Molecule X	Molecule Y
A	protein	phospholipid
B	protein	cellulose
C	phospholipid	protein
D	phospholipid	cellulose

1

3 The chart below shows the concentrations of ions in the root cells of a wheat plant and in the soil water in which it is growing.

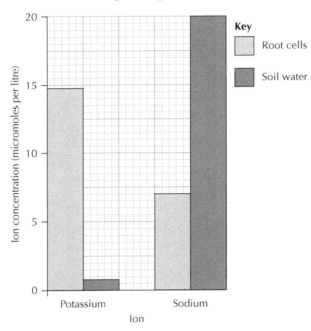

Which one of the following statements is **true**?

A Potassium ions must be taken up from soil by active transport

B Potassium ions can be taken up from soil by diffusion

C Sodium ions can pass out of the root cells by diffusion

D Sodium ions must be taken up from soil by active transport

1

4 The diagram below shows a stage in the formation of a molecule of messenger RNA (mRNA).

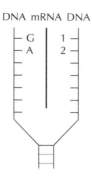

DNA mRNA DNA

G | 1
A | 2

Which line in the table below shows letters that identify correctly Bases 1 and 2?

	Base 1	Base 2
A	C	A
B	G	A
C	C	T
D	G	T

1

5 The graph below shows the effect of pH on the activity of four human digestive enzymes.

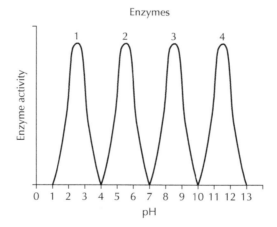

Enzymes

Which enzyme(s) work best in acid pH?

A 1 only

B 1 and 2

C 3 and 4

D 4 only

1

6 Which line in the table below correctly identifies the location of the start and the completion of the respiration pathways shown?

	Fermentation pathway		Aerobic pathway	
	starts in	completed in	starts in	completed in
A	mitochondria	cytoplasm	mitochondria	cytoplasm
B	mitochondria	mitochondria	mitochondria	cytoplasm
C	cytoplasm	cytoplasm	cytoplasm	mitochondria
D	cytoplasm	mitochondria	cytoplasm	mitochondria

1

7 Which of the following shows the fermentation pathway in animal cells?

A pyruvate → lactate

B lactate → pyruvate

C pyruvate → ethanol

D lactate → ethanol

1

8 Which row in the table below shows properties of stem cells from different locations?

	Location of stem cells			
	In the early embryo		In the adult body	
	Can self-renew	Can specialise	Can self-renew	Can specialise
A	✓	✗	✓	✓
B	✓	✓	✓	✓
C	✓	✓	✗	✓
D	✗	✓	✓	✗

1

9 The list below shows levels of organisation in the body of a mammal.

1 Organ

2 Cell

3 System

4 Tissue

Which is the correct hierarchy into which these levels can be arranged?

A 2→4→3→1

B 4→2→1→3

C 4→2→3→1

D 2→4→1→3

1

10 The flow chart below shows information about the regulation of blood glucose in humans.

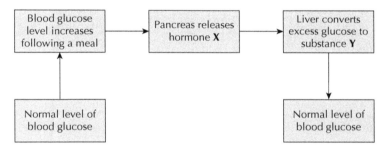

Which line in the table below identifies correctly hormone **X** and substance **Y**?

	Hormone X	Substance Y
A	insulin	starch
B	glucagon	glycogen
C	insulin	glycogen
D	glucagon	starch

1

11 The graph below shows the blood glucose concentration of a patient after he had taken a glucose drink.

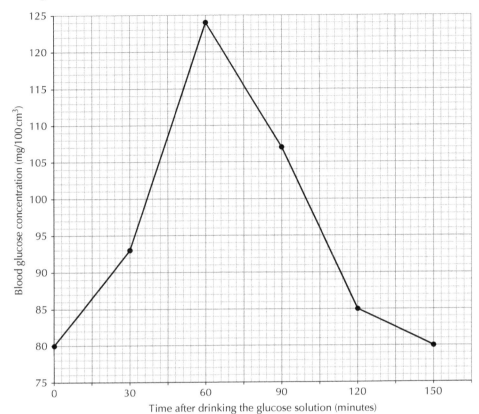

What is the percentage increase in the patient's blood glucose concentration 60 minutes after taking the drink?

A 5.5 %

B 44 %

C 55 %

D 80 %

1

12 The diagram below shows a vertical section through a flower of the pea family.

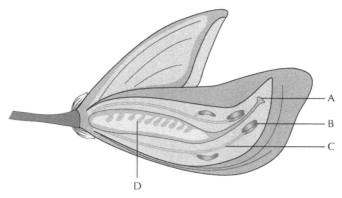

In which structure are female gametes produced?

1

13 A pea plant with yellow seeds was crossed with a pea plant with green seeds. All of the F$_1$ plants produced had yellow seeds.

The genotype of the parent plant with green seeds could be described as

A heterozygous and recessive

B homozygous and dominant

C heterozygous and dominant

D homozygous and recessive

1

Questions 14 and 15 refer to the diagrams below which show tissues from a plant.

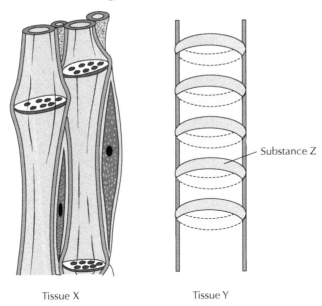

Tissue X   Substance Z

Tissue Y

14 Which line in the table below correctly identifies these tissues?

	Tissue X	*Tissue Y*
A	phloem	xylem
B	phloem	palisade
C	xylem	phloem
D	xylem	palisade

1

15 What is the function of substance Z?

A To transport water

B To withstand pressure

C To transport sugars

D To trap light

1

16 The diagram below shows part of the human digestive system.

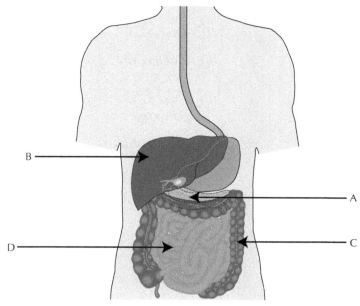

In which region of the diagram would villi be found?

17 Which term describes all the organisms living in an area and the non-living factors with which the organisms interact?

A Habitat

B Ecosystem

C Community

D Niche

18 The paired-statement key below can be used to identify duckweed plants which grow on the surface of still or slow-moving water.

1	Has roots	go to 2
	No roots	*Wolffia arrhizia*
2	Leaf flat	go to 3
	Leaf domed	*Lemna gibba*
3	Leaves pointed	*Lemna trisulca*
	Not pointed tip	go to 4
4	Many roots	*Spirodela polyrhiza*
	One root	go to 5
5	Leaves pale grey-green	*Lemna minuta*
	Leaves bright yellow-green	*Lemna minor*

Which duckweed species is shown in the drawing below?

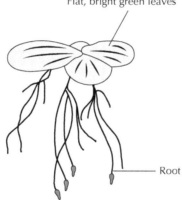

Flat, bright green leaves

Root

A   *Wolffia arrhizia*

B   *Lemna trisulca*

C   *Spirodela polyrhiza*

D   *Lemna minor*

1

19 The diagram below shows a pyramid of numbers representing a food chain.

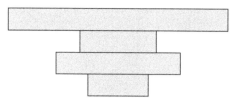

Which of the following food chains could be represented by this pyramid?

A oak tree → moth caterpillar → blue tit → feather mite (a parasite)

B oak tree → greenfly → ladybird → blue tit (a predator)

C heather → moth caterpillar → meadow pipit → merlin (a predator)

D heather → moth caterpillar → meadow pipit → feather mite (a parasite)

1

20 Some organisms living in seas off the east coast of Scotland are shown in the food web below.

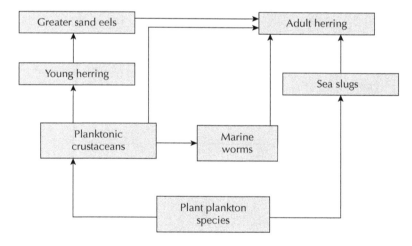

Which line in the table below correctly shows pairs of organisms that are involved in the types of competition shown?

	Type of competition	
	---	---
	interspecific	*intraspecific*
A	planktonic crustaceans and sea slugs	young and adult herring
B	young and adult herring	greater sand eels and planktonic crustaceans
C	young and adult herring	marine worms and sea slugs
D	planktonic crustaceans and sea slugs	two species of plant plankton

1

21 The diagram below shows two events in the first stage of photosynthesis in a leaf cell.

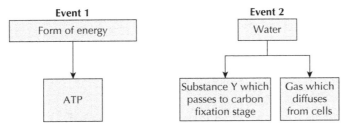

Which line in the table below correctly identifies the form of energy in event 1 and substance Y?

	Form of energy	Substance Y
A	light	oxygen
B	chemical	oxygen
C	light	hydrogen
D	chemical	hydrogen

1

22 The graph below shows the effect of increasing light intensity on the rate of photosynthesis at different temperatures and carbon dioxide concentrations.

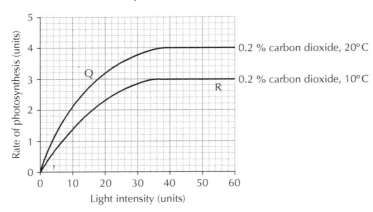

Which line in the table below correctly shows the factors that are limiting photosynthesis at points Q and R on the graph?

	Q	R
A	light intensity	temperature
B	carbon dioxide concentration	temperature
C	light intensity	carbon dioxide concentration
D	temperature	light intensity

1

23 In an investigation, the average numbers of individuals of two forms of the peppered moth in city woodland were estimated every year over a five-year period.

The results are shown on the bar chart below.

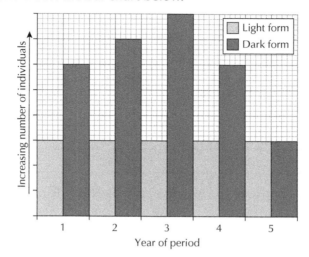

Between which two years of the period did the greatest change in the ratio of light to dark moths occur?

A 1 and 2

B 2 and 3

C 3 and 4

D 4 and 5

1

24 Plants absorb nitrates from the soil.

Which substances are produced from these nitrates by plant cells?

A Sugars

B Amino acids

C Proteins

D Phosopholipids

1

25 In an investigation, the concentration of a pesticide in the bodies of four individual birds found dead in a farmland area was measured. Two of the birds were predators, and two were prey species.

The results are shown in the table below.

Bird species	Predator or prey species	Concentration of pesticide (units per gram of muscle)
wood pigeon	prey	4
skylark	prey	2
sparrowhawk	predator	26
barn owl	predator	16

What is the difference between the average units of pesticide per gram of muscle in the prey species compared to the average in the predator species?

A 12

B 18

C 36

D 39

1

EXTENDED ANSWER QUESTIONS

 26 The diagram below represents a cell from a green plant.

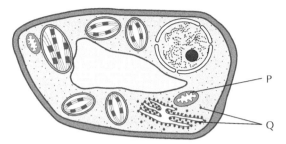

P

Q

a Give evidence from the diagram that suggests that this cell can carry out photosynthesis.

_____ 1

b Give the function of structure P.

_____ 1

c Name structures Q.

_____ 1

d Give **one** structural difference that would be expected between this cell and a fungal cell.

_____ 1

Total marks 4

27 The diagram below shows the transport of molecule S through a cell membrane.

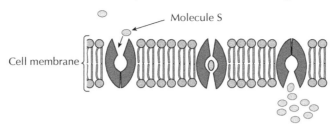

Molecule S

Cell membrane

a Name the method shown in the diagram by which molecule S is being moved across the membrane. Give **one** reason for your answer.

Method _____ **1**

Reason _____ **1**

b In an investigation, small pieces of tissue of known mass were taken from a water plant submerged in pond water. They were placed into different concentrations of sucrose solution for one hour. After this time, the mass of each piece of tissue was re-measured and expressed as a percentage of its original volume.

The results are shown in the table below.

Concentration of sucrose solution (grams per litre)	Final mass of tissue (% of its mass in pond water)
0	100.0
5	98.5
10	95.0
15	92.5
20	90.5
25	90.0

i   On the grid below, complete the vertical axis and plot a line graph to show the effect of sucrose concentration on the mass of the water plant tissue. (A spare grid, if required, can be found at the end of this section of the book.)

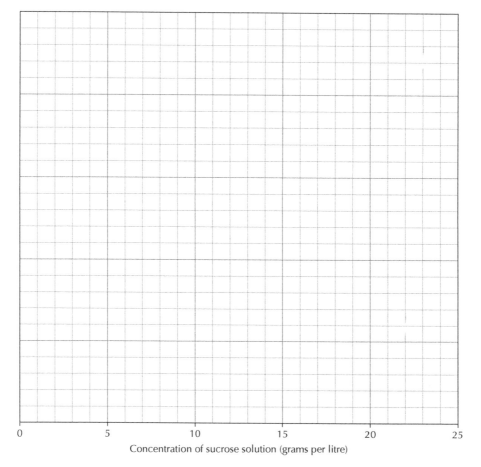

Concentration of sucrose solution (grams per litre)

2

ii Name the process that causes the mass changes in the water plant tissue.

_____

1

iii Using the information available in the table, predict the final mass of a piece of water plant tissue with a starting mass of 2.0 g after it has been immersed in a 25 % sucrose solution for one hour.

Space for calculation

Final mass = _____ g

1

Total marks 6

28 The diagram below shows the genetic modification of a bacterial cell by the transfer of a human gene.

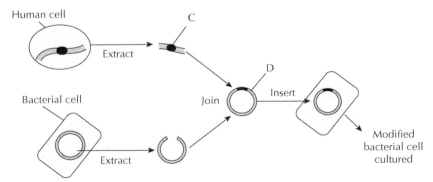

a Name the substance of which the human gene C is composed.

_____ 1

b Identify structure D.

_____ 1

c Suggest how enzymes are involved in the genetic modification process.

_____

_____ 1

d Give **one** example of a substance produced by the expression of a human gene that has been obtained by this method.

_____ 1

Total marks 4

29 When mammalian muscle tissue contracts, it decreases in length.

The diagram below shows the procedure involved in an investigation into the effect of different solutions on the lengths of pieces of mammalian muscle tissue. Each piece of muscle tissue was measured before and after five minutes of immersion in the solutions.

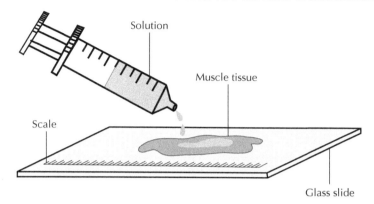

The results are shown in the table below.

Muscle tissue	Solution	Length of muscle tissue (mm)			Percentage difference in length (%)
		at start	after five minutes	difference in length	
A	1 % glucose	45	45	0	0
B	1 % ATP	50	46	4	
C	distilled water	48	48	0	0

a Complete the table by calculating the percentage decrease in length of muscle tissue B.

_____% **1**

b Explain why glucose has no effect on muscle tissue A, whereas ATP causes muscle tissue B to contract.

_____

_____ 2

c Describe why muscle tissue C was included in the experimental design.

_____

_____ 1

d State what is meant by the term **tissue** in this example.

_____ 1

Total marks 5

30 The diagrams below represent stages of mitosis in a plant cell.

Stage W   Stage X   Stage Y   Stage Z

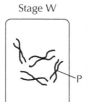   

a Name the genetic material of which structure P is composed.

_____ **1**

b Describe events which would occur at Stage Y.

_____

_____ **2**

c Give **one** use of mitosis in organisms.

_____ **1**

Total marks **4**

31 The diagram below shows some structures involved in an example of a reflex action in humans.

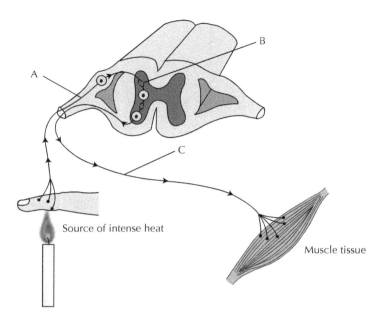

Source of intense heat

Muscle tissue

a Neurons A, B and C form the reflex arc.

Name each of these neurons.

A _____

B _____

C _____   **2**

b Identify the stimulus and describe the expected response in this example.

Stimulus _____   **1**

Description of response _____

_____   **1**

c Explain the importance of reflex actions in general.

_____

_____   **1**

Total marks **5**

32 Garden pea plants that carry the allele **T** have a tall phenotype.

Plants with the genotype **tt** are dwarf.

20 seeds of a tall variety and 20 seeds of a dwarf variety were germinated and grown for 15 weeks in a greenhouse. After this time the height of each plant was measured, and the results are shown in the charts below.

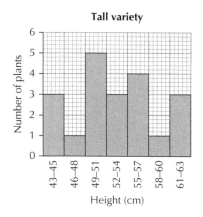

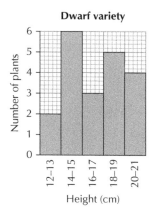

a Other than growing the same number of plants for the same time, give **two** variables that should have been kept constant to ensure that comparison of the two varieties was valid.

Variable 1 _____

1

Variable 2 _____

1

b Give the range of heights in the tall variety.

_____

1

c For the information provided, give evidence that shows that height in pea plants shows both discrete **and** continuous variation.

Evidence for discrete variation _____

_____

1

Evidence for continuous variation _____

_____

1

Total marks | **5**

33 Dimples are human facial features. Their presence is controlled by alleles of a single gene. The dominant allele (D) gives dimples and the recessive allele (d) gives no dimples.

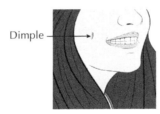

Dimple

The diagram below shows the inheritance of dimples in a family.

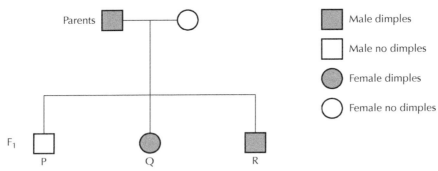

Parents

F₁

P   Q   R

Male dimples

Male no dimples

Female dimples

Female no dimples

a Give the genotypes of the following individuals.

i the female parent

_____ **1**

ii son R

_____ **1**

b i Daughter Q and a man with no dimples are expecting a baby.
Calculate the chance that their baby will inherit dimples.

Space for calculation

_____% **1**

ii Give **one** reason to explain why phenotype ratios among offspring are not always achieved.

_____ **1**

Total marks **4**

34 **a** The diagram below shows an external view of the human heart.

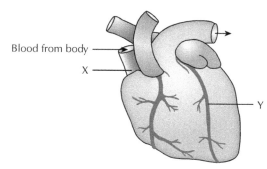

Blood from body —

X —

— Y

i Identify blood vessels X and Y.

X _____ **1**

Y _____ **1**

ii Decide if each of the statements about blood vessels in the grid below is true or false and tick (✔) the correct box.

If the statement is false write the correct word in the box to replace the word underlined in the statement.

Statement	True	False	Correction
Arteries carry blood <u>from</u> the heart.			
<u>Veins</u> exchange materials with the tissues.			
<u>Capillaries</u> have valves.			

2

b The graph below shows the effect of carbon dioxide concentration in the air on the volume of air inhaled into the lungs of an individual at rest.

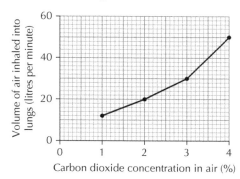

Volume of air inhaled into lungs (litres per minute)

Carbon dioxide concentration in air (%)

i Calculate the volume of carbon dioxide entering the individual's lungs each minute when the volume of air inhaled is 20 litres per minute.

Space for calculation

_____ litres **1**

ii Calculate the increase in volume of air entering the lungs per minute when the concentration of carbon dioxide in the air increases from 1 % to 4 %.

Space for calculation

_____ litres **1**

Total marks 6

35 In an investigation into the effects of temperature on the rate of transpiration in a leafy seedling, the potometer below was set up.

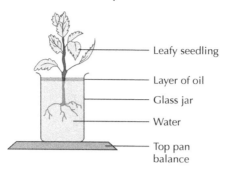

- Leafy seedling
- Layer of oil
- Glass jar
- Water
- Top pan balance

Transpiration rate was measured at different temperatures. The results are shown in the table below.

Temperature (°C)	Transpiration rate (grams of water per cm² of leaf per minute)
10	0.2
15	0.3
20	0.4
25	0.5

a Identify the observations or measurements that would have to be made to obtain the values for the rate of transpiration shown in the table.

_____

_____

_____ **3**

b The following factors can affect transpiration rate in plants.

Light intensity **Atmospheric humidity** **Air movements**

Choose **one** of these factors and describe how the apparatus above could be modified to investigate this factor.

Factor _____

Description _____

_____ **2**

Total marks **5**

36 The pie chart below shows one estimate of the percentage of the Earth's land area occupied by different ecosystems.

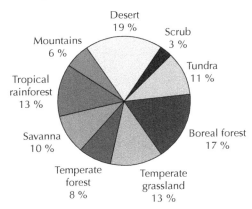

a Desert ecosystems are found in many parts of the world.

   i Apart from the organisms which live there, state what else makes up a desert ecosystem.

     _____   **1**

   ii Desert organisms have adaptations for their way of life.

      Describe what is meant by an adaptation.

     _____

     _____   **1**

b Calculate the total percentage of land surface occupied by the various types of forest ecosystems.

Space for calculation

                                                     _____%   **1**

c Within ecosystems, each species occupies its own niche.

   Describe what is meant by the term niche.

     _____

     _____   **1**

d The diagram shows a food chain from a temperate forest ecosystem.

The arrows represent energy flow.

oak tree leaves ⟶ greenfly ⟶ ladybird ⟶ blue tit ⟶ sparrowhawk

    i Explain how the oak tree leaves obtain the energy trapped within them.

_____

_____ 1

    ii Give **one** fate of energy present in the greenfly population which does not pass to the ladybird population.

_____ 1

Total marks 6

37 In an investigation to compare the populations of a species of ground beetle living on the soil surface in two different areas of grassland, sampling was carried out using pitfall traps.

a Give **two** precautions that would have to be taken to ensure that the sampling method allowed valid **comparison** of the two areas.

1 _____

2 _____ 2

b Describe a source of error that can arise when using pitfall traps.

_____ 1

c During the investigation, a number of abiotic factors related to the soil were also measured.

Name **one** abiotic factor that is related to soil and describe how it could be measured.

Abiotic factor _____ 1

Method of measurement _____

_____ 1

Total marks 5

38　**a**　The apparatus shown below was set up to measure the rate of photosynthesis in pondweed.

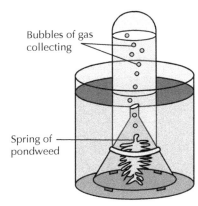

Bubbles of gas collecting

Spring of pondweed

Lamp

i　Name the gas present in the collecting bubbles.

_____

1

ii　Describe how the apparatus could be used to show the effects of light intensity on the rate of photosynthesis.

_____

_____

_____

3

b　The graph below shows some results of an experiment to show the effect of carbon dioxide concentration on the rate of photosynthesis at different temperatures.

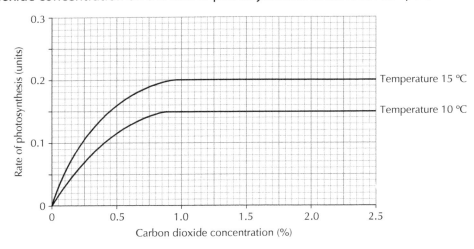

Temperature 15 °C

Temperature 10 °C

i   Give the carbon dioxide concentration at which the rate of photosynthesis first
    reached its maximum at 15° C.

_____% **1**

ii Calculate the increase in the rate of photosynthesis when the temperature was
    raised from 10° C to 15° C at a carbon dioxide concentration of 2.0 %.

Space for calculation

_____units **1**

c   Sugar produced by photosynthesis can be converted into starch.
    Give the function of starch in green plants.

_____ **1**

Total marks **7**

39 Read the following passage and answer the questions based on it.

Genetically modified (GM) rice

Nitrogen is the most important soil nutrient for plants and a major factor which can limit crop productivity. Nitrogen-rich fertilisers are often used to boost crop growth but plants are inefficient at taking up the nitrate from applied fertiliser. As a result, excess nitrates frequently leach from the soil into waterways and cause algal blooms. Dead algae become food for bacteria which consume oxygen needed for healthy aquatic ecosystems.

To meet growing food demands, the global use of nitrate fertiliser increased from 3.5 million metric tonnes in 1960 to 87 million metric tonnes in 2000, and is projected to increase to 249 million metric tonnes by the year 2050.

Scientists in Canada have successfully developed a genetically modified (GM) rice variety which produces enzymes that allow more efficient nitrate uptake than unmodified varieties. This may reduce the need for nitrogen-rich fertilisers and at the same time increase yields.

These crops not only have the potential to lower production costs and reduce environmental pollution, but their increased productivity could make a significant contribution to our long-term food security.

a Explain why fertiliser use creates excess nitrates in soils.

_____ **1**

b Explain how algal blooms can lead to the de-oxygenation of freshwater ecosystems.

_____ **1**

c Calculate the projected global increase in nitrate fertiliser use between 1960 and 2050.

Space for calculation

_____ metric tonnes **1**

d Explain why the GM rice described in the passage could reduce the amount of nitrate fertiliser which needs to be applied.

_____ **1**

e Apart from reducing fertiliser application, give **one** other benefits of growing the GM rice.

_____ **1**

Total marks 5

40 On the Galapagos Islands of the Pacific Ocean, speciation has produced a group of similar finch species, as shown in the diagram below. The group arose from a single ancestor species, which reached the islands from the South American mainland millions of years ago.

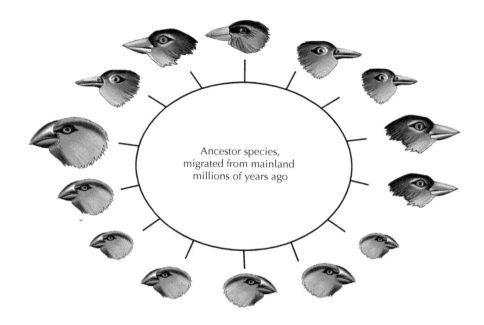

Ancestor species, migrated from mainland millions of years ago

a The list below shows processes involved in speciation.

mutation   isolation   natural selection

Describe how these processes have led to the production of the group of finch species in the diagram above.

_____

_____

_____

_____

_____ 3

b Give the term applied to mutations that confer neither advantage nor disadvantage.

_____ 1

Total marks 4

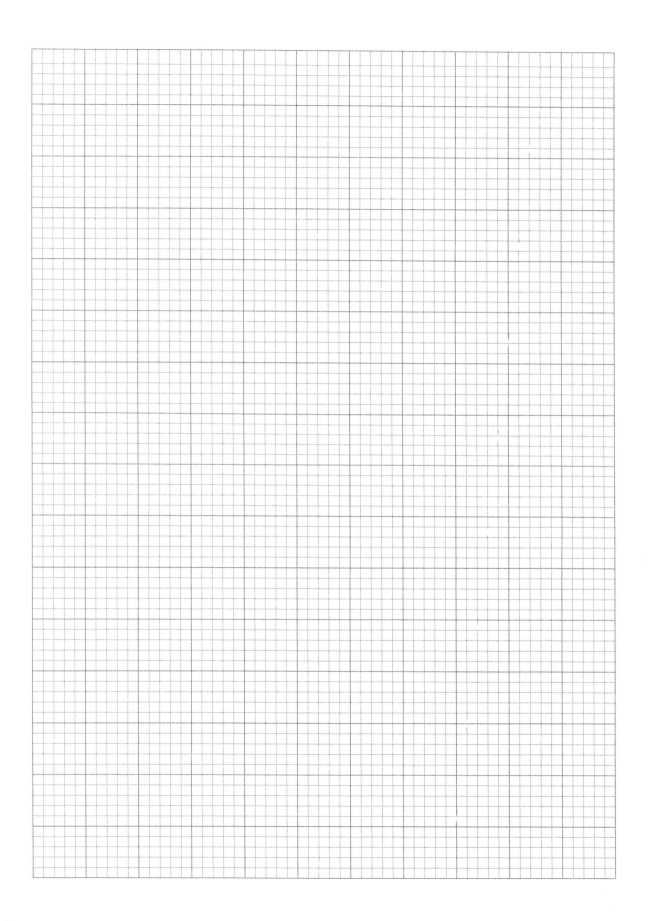

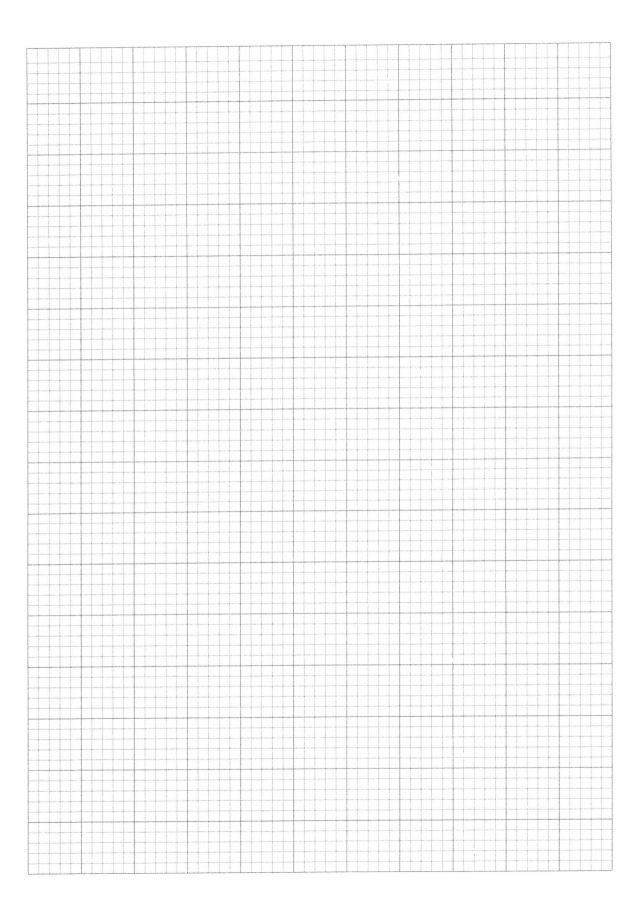

Notes

Notes